T0140270

Studies in Systems, Decision and Control

Volume 62

Series editor

Janusz Kacprzyk, Polish Academy of Sciences, Warsaw, Poland
e-mail: kacprzyk@ibspan.waw.pl

About this Series

The series "Studies in Systems, Decision and Control" (SSDC) covers both new developments and advances, as well as the state of the art, in the various areas of broadly perceived systems, decision making and control- quickly, up to date and with a high quality. The intent is to cover the theory, applications, and perspectives on the state of the art and future developments relevant to systems, decision making, control, complex processes and related areas, as embedded in the fields of engineering, computer science, physics, economics, social and life sciences, as well as the paradigms and methodologies behind them. The series contains monographs, textbooks, lecture notes and edited volumes in systems, decision making and control spanning the areas of Cyber-Physical Systems, Autonomous Systems, Sensor Networks, Control Systems, Energy Systems, Automotive Systems, Biological Systems, Vehicular Networking and Connected Vehicles, Aerospace Systems, Automation, Manufacturing, Smart Grids, Nonlinear Systems, Power Systems, Robotics, Social Systems, Economic Systems and other. Of particular value to both the contributors and the readership are the short publication timeframe and the world-wide distribution and exposure which enable both a wide and rapid dissemination of research output.

More information about this series at http://www.springer.com/series/13304

Emil Pricop · Grigore Stamatescu
Editors

Recent Advances in Systems Safety and Security

 Springer

Editors
Emil Pricop
Department of Automatic Control,
  Computers and Electronics
Petroleum-Gas University of Ploieşti
Ploieşti
Romania

Grigore Stamatescu
Department of Automatic Control
  and Industrial Informatics
University Politehnica of Bucharest
Bucharest
Romania

ISSN 2198-4182   ISSN 2198-4190 (electronic)
Studies in Systems, Decision and Control
ISBN 978-3-319-81308-0   ISBN 978-3-319-32525-5 (eBook)
DOI 10.1007/978-3-319-32525-5

Printed on acid-free paper

This Springer imprint is published by Springer Nature
The registered company is Springer International Publishing AG Switzerland

Foreword

Safety is "the condition of being protected from or unlikely to cause danger, risk, or injury". Security is "the state of being free from danger or threat".

As the emergence of networked intelligent systems is being witnessed in multiple areas of technical applications, ranging from industrial control systems, embedded solutions for the transportation domain, border protection and access control, pervasive instrumentation for urban environments and others, fundamental research and engineering challenges related to their safety and security have been addressed. This continues to occur under a rapid pace of research and development.

Theoretical approaches conventionally cover the conceptual and abstract modeling of threat scenarios and fallback solutions, increasing the overall resilience of the developed systems. Mathematical modeling and proofs thus offer guarantees and bounds to clearly highlight the limitations under the state of the art. In parallel, applications are concerned mainly with implementing novel technologies and solutions across domains, which can include but are not limited to wireless sensor networks, distributed controllers, smart and connected cameras or various robotic systems. Also, the software frameworks supporting the deployment of safe and secure systems have started to become modular and distributed, while even offloading some components to the cloud. This brings upon added complexity and the need to protect the systems at every end point and to make sure that the interfaces that enable the integration with other systems are soundly designed.

The International Workshop on Systems Safety and Security (IWSSS, http://www.iwsss.org), in 2016 at its fourth edition, was initially started with the idea of bringing together specialists interested in these topics. It has since become an annual platform for exchange of experience and ideas which intends to stimulate joint work at a regional and international level.

The main goal of the current book was thus to collect, as chapters and under a unifying direction, selected contributions that represent original chapters based on extended works from the 2nd—2014 and 3rd—2015 editions of IWSSS, held in Bucharest, Romania. Its contributions are aimed at both scientists engineers working in this field and represent a timely overview of advances in systems safety and security. The book structure and chapters are broadly grouped into core topics and address challenges related to information theoretic methods for assuring systems safety and security, cloud-based solutions, image processing approaches, distributed sensor networks and legal or risk analysis viewpoints. These are mostly accompanied by associated case studies, providing additional practical value and underlying the broad relevance and impact of the field.

We hope the book represents a useful reference and tool to a broad base of professionals in research and industry alike and to further foster cooperation in this area of critical interest.

April 2016   The Editors

Contents

Contributors

Dorin Carstoiu Department of Automatic Control and Industrial Informatics, University Politehnica of Bucharest, Bucharest, Romania

Alexandra Cernian Department of Automatic Control and Industrial Informatics, University Politehnica of Bucharest, Bucharest, Romania

Jun Chen Idaho National Laboratory, Idaho Falls, USA

Oana Chenaru Department of Automatic Control and Industrial Informatics, Faculty of Automatic Control and Computer Science, University Politehnica of Bucharest, Bucharest, Romania

Viktor M. Denisov Flagman Geo Ltd., Saint-Petersburg, Russia

Radu Dobrescu Department of Automatic Control and Industrial Informatics, University Politehnica of Bucharest, Bucharest, Romania

Jaouhar Fattahi Département d'informatique et de génie logiciel Pavillon Adrien-Pouliot, Université Laval, Quebec, QC, Canada

Radu Fratila Faculty of Automatic Control and Computers, University Politehnica of Bucharest, Bucharest, Romania

Octavian Mihai Ghita Department of Electrical Measurements, Faculty of Electrical Engineering, University Politehnica of Bucharest, Bucharest, Romania

Diana Gornea Faculty of Automatic Control and Computers, University Politehnica of Bucharest, Bucharest, Romania

Ionela Halcu Intelligent Measurement Technologies and Transducers Laboratory, University Politehnica of Bucharest, Bucharest, Romania

Ján Host Faculty of Science, Institute of Computer Science, Pavol Jozef Šafárik University in Košice, Košice, Slovakia

Mariam Ibrahim Department of Electrical and Computer Engineering, Iowa State University, Ames, IA, USA; Department of Mechatronics Engineering, German Jordanian University, Amman, Jordan

Loretta Ichim Faculty of Automatic Control and Computers, University Politehnica of Bucharest, Bucharest, Romania; "Stefan S. Nicolau" Institute of Virology, Bucharest, Romania

Ratnesh Kumar Department of Electrical and Computer Engineering, Iowa State University, Ames, IA, USA

Yutaka Matsubara Nagoya University, Nagoya, Aichi, Japan

Mohamed Mejri Département d'informatique et de génie logiciel Pavillon Adrien-Pouliot, Université Laval, Quebec, QC, Canada

Sanda Florentina Mihalache Department of Automatic Control, Computers and Electronics, Petroleum-Gas University of Ploieşti, Ploieşti, Prahova, Romania

Dorel Nasui Department of Automatic Control and Industrial Informatics, University Politehnica of Bucharest, Bucharest, Romania

Marilena Nicolae Petroleum Processing Engineering and Environmental Protection Department, Petroleum-Gas University of Ploieşti, Ploieşti, Romania

Dan Popescu Department of Automatic Control and Industrial Informatics, Faculty of Automatic Control and Computer Science, University Politehnica of Bucharest, Bucharest, Romania

Emil Pricop Department of Automatic Control, Computers and Electronics, Petroleum-Gas University of Ploieşti, Ploieşti, Prahova, Romania

Gabriel Rădulescu Department of Automatic Control, Computers and Electronics, Petroleum-Gas University of Ploieşti, Ploieşti, Romania

Cosmina Roşca Department of Automatic Control, Computers and Electronics, Petroleum-Gas University of Ploieşti, Ploieşti, Romania

Masaki Samejima Graduate School of Information Science and Technology, Osaka University, Suita-shi, Osaka, Japan

Naoki Satoh Wakayama University, Wakayama-shi, Wakayama, Japan

Paul Schiopu Faculty of Electronics, Telecommunications and Information Technology, University Politehnica of Bucharest, Bucharest, Romania; Department of Electronic Technology and Reliability, University Politehnica of Bucharest, Bucharest, Romania

Valentin Sgârciu Department of Automatic Control and Industrial Informatics, Faculty of Automatic Control and Computers, University Politehnica of Bucharest, Bucharest, Romania

Pavol Sokol Faculty of Science, Institute of Computer Science, Pavol Jozef Šafárik University in Košice, Košice, Slovakia

Grigore Stamatescu Department of Automatic Control and Industrial Informatics, Faculty of Automatic Control and Computers, University Politehnica of Bucharest, Bucharest, Romania

Iulia Stamatescu Department of Automatic Control and Industrial Informatics, Faculty of Automatic Control and Computers, University Politehnica of Bucharest, Bucharest, Romania

Hiroaki Takada Nagoya University, Nagoya, Aichi, Japan

Andrey V. Timofeev LLP EqualiZoom, Astana, Kazakhstan

Adela Vintea Department of Electronic Technology and Reliability, Faculty of Electronics, Telecommunications and Information Technology, University Politehnica of Bucharest, Bucharest, Romania

Jingxuan Wei Nagoya University, Nagoya, Aichi, Japan

The Theory of Witness-Functions

Jaouhar Fattahi, Mohamed Mejri and Emil Pricop

Abstract Cryptographic protocols are distributed programs that ensure security in all communications. They guarantee agents authentication, data confidentiality, data integrity, atomicity of goods and money, non-repudiation, etc. They are used in all areas: e-commerce, military fields, electronic voting, etc. The use of cryptography is essential to ensure protocols' security, however, it is not sufficient. Indeed, in the literature, a significant number of cryptographic protocols have long been considered safe, but they were shown faulty many years after their use. Saying that a protocol is correct or not is an undecidable problem in general. However, several methods (logic-based methods, Model-Checking-based methods, typing-based methods, etc.) have emerged to answer this hard question under restrictive assumptions and led to varying results. Here, we present a new formal method to analyze cryptographic protocols statically for the property of secrecy. It consists in inspecting the level of security of every component of exchanged messages in the protocol by new metrics, called witness-functions, and making sure that it does not diminish during its life cycle. If yes, we declare that the protocol keeps its secret inputs. We analyze here an amended version of the Woo-Lam protocol using the witness-functions' theory.

J. Fattahi (✉) · M. Mejri
Département d'informatique et de génie logiciel Pavillon Adrien-pouliot,
Université Laval, 1065, av. de la Médecine, Quebec, QC G1V 0A6, Canada
e-mail: jaouhar.fattahi.1@ulaval.ca

M. Mejri
e-mail: mohamed.mejri@ift.ulaval.ca

E. Pricop
Petroleum-Gas University of Ploieşti, Ploieşti, Romania
e-mail: emil.pricop@upg-ploiesti.ro

© Springer International Publishing Switzerland 2016
E. Pricop and G. Stamatescu (eds.), *Recent Advances in Systems Safety and Security*,
Studies in Systems, Decision and Control 62, DOI 10.1007/978-3-319-32525-5_1

1

1 Introduction

In this work, we present the witness-functions as a new formal method for analyzing protocols and we run an analysis on an amended version of the Woo-Lam protocol using one of them. The witness-functions have been recently introduced by Fattahi et al. [1–6] to statically analyze cryptographic protocols for secrecy. A protocol analysis with a witness-function consists in inspecting every component in the protocol in order to make sure that its security never drops between any receiving step and a subsequent sending one. If yes, the protocol is said to be increasing and we conclude that it keeps its secret inputs. We use the witness-function to evaluate the security of every component in the protocol.

This work is organized as follows:

- First, we give some notations that we will use in this work;
- then, in the Sect. 2, we give some abstract conditions on a function to be safe for a protocol analysis and we state that an increasing protocol keeps its secret inputs when analyzed using such functions;
- then, in the Sects. 3 and 4, we present the witness-functions and we highlight their advantages, particularly their static bounds. We state the theorem of protocol analysis with the witness-functions, as well;
- then, in the Sect. 5, we run an analysis on an amended version of the Woo-Lam protocol and we interpret the results;
- finally, we compare our witness-functions with some related works and we conclude.

1.1 Notations

Here, we give some notations and conventions that will be used throughout this work.

- We denote by $\mathscr{C} = \langle \mathscr{M}, \xi, \vDash, \mathscr{K}, \mathscr{L}^{\beth}, \ulcorner \cdot \urcorner \rangle$ the context containing the parameters that affect the analysis of a protocol:

  - $\mathscr{M}$: is a set of messages built from the algebraic signature $\langle \mathscr{N}, \Sigma \rangle$ where $\mathscr{N}$ is a set of atomic names (nonces, keys, principals, etc.) and Σ is a set of functions (*enc*:: encryption, *dec*:: decryption, *pair*:: concatenation (denoted by ".." here), etc.). I.e. $\mathscr{M} = T_{\langle \mathscr{N}, \Sigma \rangle}(\mathscr{X})$. We use Γ to denote the set of all substitutions from $\mathscr{X} \to \mathscr{M}$. We designate by $\mathscr{A}$ all atomic messages (atoms) in $\mathscr{M}$, by $\mathscr{A}(m)$ the set of atomic messages in m and by $\mathscr{I}$ the set of principals including the intruder I. We denote by k^{-1} the reverse key of a key k and we consider that $(k^{-1})^{-1} = k$.
  - ξ: is the theory that describes the algebraic properties of the functions in Σ by equations. e.g. $dec(enc(x, y), y^{-1}) = x$.

- $\vDash$:
- $\vDash_\mathscr{C}$: is the inference system of the intruder under the equational theory ξ. Let M be a set of messages and m be a message. $M \vDash_\mathscr{C} m$ expresses that the intruder can infer m from M using her capacity. We extend this notation to valid traces as follows: $\rho \vDash_\mathscr{C} m$ means that the intruder can deduce m from the messages in the trace ρ. We assume that the intruder has the full control of the net as described in the Dolev-Yao model [7]. She may redirect, delete and modify any message. She holds the public keys of all participants, her private keys and the keys that she shares with other participants. She can encrypt or decrypt any message with the keys that she holds. Formally, the intruder has generically the following rules for building messages:

$$(init) : \frac{\Box}{M \vDash_\mathscr{C} m} [m \in M \cup K(I)]$$

$$(op) : \frac{M \vDash_\mathscr{C} m_1, \ldots, M \vDash_\mathscr{C} m_n}{M \vDash_\mathscr{C} f(m_1, \ldots, m_n)} [f \in \Sigma]$$

$$(eq) : \frac{M \vDash_\mathscr{C} m', m' =_\mathscr{C} m}{M \vDash_\mathscr{C} m}, \quad \text{with } (m' =_\mathscr{C} m \equiv m' =_{\xi(\mathscr{C})} m)$$

Example 1 The intruder capacity can be described by the following rules:

$$(init) : \frac{\Box}{M \vDash_\mathscr{C} m} [m \in M \cup K(I)]$$

$$(dec) : \frac{M \vDash_\mathscr{C} k, M \vDash_\mathscr{C} m_k}{M \vDash_\mathscr{C} m}$$

$$(enc) : \frac{M \vDash_\mathscr{C} k, M \vDash_\mathscr{C} m}{M \vDash_\mathscr{C} \{m\}_k}$$

$$(concat) : \frac{M \vDash_\mathscr{C} m_1, M \vDash_\mathscr{C} m_2}{M \vDash_\mathscr{C} m_1 \cdot m_2}$$

$$(deconcat) : \frac{M \vDash_\mathscr{C} m_1 \cdot m_2}{M \vDash_\mathscr{C} m_i} [i \in \{1, 2\}]$$

In this example, from a set of messages, an intruder can infer any message in this set, encrypt any message when she possesses previously the encryption key, decrypt any message when she possesses previously the decryption key, concatenate any two messages and deconcatenate them.

- $\mathscr{K}$: is a function from $\mathscr{I}$ to $\mathscr{M}$, that assigns to any principal a set of atomic messages describing her initial knowledge. We denote by $K_\mathscr{C}(I)$ the initial knowledge of the intruder, or simply $K(I)$ where the context is obvious.
- $\mathscr{L}^\sqsupseteq$: is the security lattice $(\mathscr{L}, \sqsupseteq, \sqcup, \sqcap, \bot, \top)$ used to assign security values to messages. A concrete example of a lattice is $(2^\mathscr{I}, \subseteq, \cap, \cup, \mathscr{I}, \emptyset)$ that will be used in this work.

– $\ulcorner \cdot \urcorner$: is a partial function that assigns a value of security (type) to a message in $\mathcal{M}$. Let M be a set of messages and m a single message. We write $\ulcorner M \urcorner \sqsupseteq \ulcorner m \urcorner$ when $\exists m' \in M \cdot \ulcorner m' \urcorner \sqsupseteq \ulcorner m \urcorner$.

• Our analysis takes place in a role-based specification. A role-based specification is a finite set of generalized roles. A generalized role is a protocol abstraction where the emphasis is put on some principal and where all the unknown messages, that could not be verified, are replaced by variables. An exponent i (the session identifier) is added to a fresh message to express that this component changes values from one execution to another. A generalized role expresses how an agent sees and understands the exchanged messages. A generalized role may be extracted from a protocol by these steps:

1. We extract the roles from a protocol.
2. We replace the unknown messages by fresh variables in each role.

Roles can be extracted following these steps:

1. For every principal, we extract all the steps in which she participates. Then, we add a session identifier i in the steps identifiers and in fresh values.
   For example, from the Woo-Lam protocol given in Table 1, we extract three roles, denoted by R_A (for the agent A), R_B (for the agent B), and R_S (for the server S).
2. We introduce an intruder I to express the fact that received messages and sent messages are seemingly received or sent by an intruder.
3. Finally, we extract all prefixes from these roles. A prefix must end by a sending step.

From the roles, we define the generalized roles. A generalized role is an abstraction of a role where unknown messages are converted to variables. Indeed, a message or a component of a message is replaced by a variable when the receiver cannot make any verification on it and so she is not sure about its integrity or its origin. The generalized roles give an accurate idea on the behavior and the knowledge of participants during the protocol execution. The generalized roles of A are:

Table 1 The Woo-Lam Protocol (original version)

$$
\begin{aligned}
p := \; & \langle 1, A \rightarrow B : A \rangle. \\
& \langle 2, B \rightarrow A : N_b \rangle. \\
& \langle 3, A \rightarrow B : \{N_b.k_{ab}\}_{k_{as}} \rangle. \\
& \langle 4, B \rightarrow S : \{A.\{N_b.k_{ab}\}_{k_{as}}\}_{k_{bs}} \rangle. \\
& \langle 5, S \rightarrow B : \{N_b.k_{ab}\}_{k_{bs}} \rangle.
\end{aligned}
$$

$$\begin{aligned}
\mathscr{A}_G^1 &= \langle i.1, & A & \rightarrow & I(B) & : & A\rangle \\
\mathscr{A}_G^2 &= \langle i.1, & A & \rightarrow & I(B) & : & A\rangle. \\
& \langle i.2, & I(B) & \rightarrow & A & : & X\rangle. \\
& \langle i.3, & A & \rightarrow & I(B) & : & \{X.k_{ab}^i\}_{k_{as}}\rangle
\end{aligned}$$

The generalized roles of B are:

$$\begin{aligned}
\mathscr{B}_G^1 &= \langle i.1, & I(A) & \rightarrow & B & : & A\rangle. \\
& \langle i.2, & B & \rightarrow & I(A) & : & N_b\rangle \\
\mathscr{B}_G^2 &= \langle i.1, & I(A) & \rightarrow & B & : & A\rangle. \\
& \langle i.2, & B & \rightarrow & I(A) & : & N_b\rangle. \\
& \langle i.3, & I(A) & \rightarrow & B & : & Y\rangle. \\
& \langle i.4, & B & \rightarrow & I(S) & : & \{A.Y\}_{k_{bs}}\rangle \\
\mathscr{B}_G^3 &= \langle i.1, & I(A) & \rightarrow & B & : & A\rangle. \\
& \langle i.2, & B & \rightarrow & I(A) & : & N_b\rangle. \\
& \langle i.3, & I(A) & \rightarrow & B & : & Y\rangle. \\
& \langle i.4, & B & \rightarrow & I(S) & : & \{A.Y\}_{k_{bs}}\rangle. \\
& \langle i.5, & I(S) & \rightarrow & B & : & \{N_b^i.Z\}_{k_{bs}}\rangle
\end{aligned}$$

The generalized role of S is:

$$\begin{aligned}
\mathscr{S}_G^1 &= \langle i.4, & I(B) & \rightarrow & S & : & \{A, \{U, V\}_{k_{as}}\}_{k_{bs}}\rangle. \\
& \langle i.5, & S & \rightarrow & I(B) & : & \{U.V\}_{k_{bs}}\rangle
\end{aligned}$$

Hence, the role-based specification of the protocol described by Table 1 is $\mathscr{R}_G(p) = \{\mathscr{A}_G^1, \mathscr{A}_G^2, \mathscr{B}_G^1, \mathscr{B}_G^2, \mathscr{B}_G^3, \mathscr{S}_G^1\}$. The role-based specification is a model to formalize the concept of valid traces of a protocol. More details about the role-based specification are in [8–10].

- We denote by $\mathscr{M}_p^{\mathscr{G}}$ the set of messages (closed and with variables) generated by $R_G(p)$, by $\mathscr{M}_p$ the set of closed messages generated by substitution in terms in $\mathscr{M}_p^{\mathscr{G}}$. We denote by R^- (respectively R^+) the set of received messages (respectively sent messages) by a principal in the role R. Conventionally, we use uppercases for sets or sequences and lowercases for single elements. For example M denotes a set of messages, m a message, R a role composed of sequence of steps, r a step and $R.r$ the role ending by the step r.
- A valid trace is a close message obtained by substitution in the generalized roles. We denote by $\llbracket p \rrbracket$ the infinite set of valid traces of p.

2 An Increasing Protocol Keeps Its Secret Inputs

Hereafter, we give two abstract conditions on a function to be good for verification (safe). Then, we enunciate that an increasing protocol keeps its secret inputs.

2.1   Safe Functions

Definition 1 (*Well-built Function*) Let F be a function and $\mathscr{C}$ be a context. F is $\mathscr{C}$-well-built iff:

$$\forall M, M_1, M_2 \subseteq \mathscr{M}, \forall \lambda \in \mathscr{A}(M) : \begin{cases} F(\lambda, \{\lambda\}) = \bot; \\ F(\lambda, M_1 \cup M_2) = F(\lambda, M_1) \sqcap F(\lambda, M_2); \\ F(\lambda, M) = \top, \ if \ \lambda \notin \mathscr{A}(M). \end{cases}$$

A well-built function F must return the infimum for an atom λ that appears in clear in M to express the fact that it is exposed to everybody in M. It should return for it in the union of two sets, the minimum of the two values evaluated in each set apart. It returns the supremum for any atom λ that does appear in M to express the fact that none could deduce it from M.

Definition 2 (*Invariant-by-Intruder Function*) Let F be a function and $\mathscr{C}$ be a context. F is $\mathscr{C}$-invariant-by-intruder iff:

$$\forall M \subseteq \mathscr{M}, m \in \mathscr{M} \cdot M \vDash_{\mathscr{C}} m \Rightarrow \forall \lambda \in \mathscr{A}(m) \cdot (F(\lambda, m) \sqsupseteq F(\lambda, M)) \vee (\ulcorner K(I) \urcorner \sqsupseteq \ulcorner \lambda \urcorner).$$

An invariant-by-intruder function F is such that, when it assigns a security value to an atom λ in a set of messages M the intruder can never deduce, using her knowledge, from M another message m in which this value decreases (i.e. $F(\lambda, m) \not\sqsupseteq F(\lambda, M)$), except when λ is intentionally destined to the intruder (i.e. $\ulcorner K(I) \urcorner \sqsupseteq \ulcorner \lambda \urcorner$).

Definition 3 (*Safe Function*) Let F be a function and $\mathscr{C}$ be a context.

$$F \ is \ \mathscr{C}\text{-}safe \ iff \begin{cases} F \ is \ \mathscr{C}\text{-}well\text{-}built \\ F \ is \ \mathscr{C}\text{-}invariant\text{-}by\text{-}intruder \end{cases}$$

A safe function F is well-built and invariant-by-intruder.

Definition 4 (*F-Increasing Protocol*) Let F be a function, $\mathscr{C}$ be a context and p be a protocol.

p is F-increasing in $\mathscr{C}$ iff:
$$\forall R.r \in R_G(p), \forall \sigma \in \Gamma : \mathscr{X} \to \mathscr{M}_p \text{ we have:}$$

$$\forall \lambda \in \mathscr{A}(M).F(\lambda, r^+ \sigma) \sqsupseteq \ulcorner \lambda \urcorner \sqcap F(\lambda, R^- \sigma)$$

An F-increasing protocol generates permanently traces with atomic messages having always a security value, evaluated by F, higher when sending (i.e. in $r^+ \sigma$) than it was on its reception (i.e. in $R^- \sigma$).

Theorem 1 (Security of Increasing Protocols) *Let F be a $\mathscr{C}$-safe Function and p an F-increasing protocol.*

$$p \text{ keeps its secret inputs.}$$

Theorem 1 states that a protocol is secure when verified by a safe function F on which it is proved increasing. That is, if the intruder manages to infer a secret λ (get it in clear), then its value returned by F is the infimum because F is well-built. That could not happen due to the protocol rules because the protocol is increasing by F unless λ has initially the infimum. In this case, λ was not from the beginning a secret. That could not happen neither by using the capacity of the intruder because F is invariant-by-intruder. Therefore, the secret is kept forever.

3   Safe Functions

Now, we define three practicalfunctions that meet the conditions of safety: F_{MAX}^{EK}, F_N^{EK} and F_{EK}^{EK}. Each function among them returns for an atom λ in a message m:

1. if λ is encrypted by a key k, where k is the most external protective key (shortly the external protective key denoted by EK) that satisfies: $\ulcorner k^{-1} \urcorner \sqsupseteq \ulcorner \lambda \urcorner$, *any subset* among the principals that know k^{-1} and the principals that travel with λ under the same protection by k. At this step:

   (a) F_{MAX}^{EK} returns the set of all these candidates;
   (b) F_N^{EK} returns the set of principals that travel with λ under the same protection by k;
   (c) F_{EK}^{EK} returns the set of principals that know k^{-1}.

2. for two messages linked by an operator other than an encryption by a protective key (e.g. pair), the union of two values evaluated in the two messages apart by F.
3. if λ does not have a protective key in m, the infimum to express the fact that it could be discovered by an intruder from m;
4. if λ does not appear in m, the supremum to reflect that it could not be discovered by anybody from m;

A such function is well-built by construction. It is invariant-by-intruder too. The main idea of its invariance by intruder property is that the returned candidates (principals) are selected from a section (a component of m) protected by k (invariant by intruder). Hence, to alter this section (to lower the value of security of an atom λ), the intruder must previously have got the atomic key k^{-1}, so her knowledge should satisfy: $\ulcorner K(I) \urcorner \sqsupseteq \ulcorner k^{-1} \urcorner$. Since the key k^{-1} must satisfy: $\ulcorner k^{-1} \urcorner \sqsupseteq \ulcorner \lambda \urcorner$, then the knowledge of the intruder satisfy: $\ulcorner K(I) \urcorner \sqsupseteq \ulcorner \lambda \urcorner$ too (transitivity of "$\sqsupseteq$" in the lattice), which is the definition of an invariant-by-intruder function. It is very important to mention that we consider the form $m_\downarrow$ of a message m that removes keys that cancel out (i.e. $dec(enc(m, k), k^{-1})_{\downarrow} = m$). We suppose that we do not have any other special algebraic properties in the equational theory. This will be the scope of a future work.

Example 2 Let λ be an atom, m be a message and k_{ab} be a key such that:

$$\ulcorner \lambda \urcorner = \{A, B, S\}; m = \{A.\{S.\lambda.D\}_{k_{as}}\}_{k_{ab}}; \ulcorner k_{ab}^{-1} \urcorner = \{A, B\}$$
$$F_{MAX}^{EK}(\lambda, m) = \ulcorner k_{ab}^{-1} \urcorner \cup \{A, S, D\} = \{A, B\} \cup \{A, S, D\} = \{A, B, S, D\}.$$
$$F_{MAX}^{N}(\lambda, m) = \{A, S, D\}.$$
$$F_{MAX}^{EK}(\lambda, m) = \ulcorner k_{ab}^{-1} \urcorner = \{A, B\}.$$

In the rest of this work, F refers to any of the functions F_{MAX}^{EK}, F_N^{EK} and F_{EK}^{EK}.

4   The Witness-Functions

According to Theorem 1, if a protocol p is proved F-increasing on its *valid traces* using a safe function F, then it is secure. However, the set of valid traces is infinite. In order to be able to analyze a protocol from within its finite set of the generalized roles, we should adapt a safe function to the problem of substitution (variables) and look for an additional mechanism that allows us to propagate any decision made on the generalized roles to valid traces. The witness-functions are this mechanism. But first, let us introduce the derivative messages. A derivative message is a message of the generalized roles from which we exclude variables that do not contribute to the evaluation of security. This is described in Definition 5.

Definition 5 (*Derivation*) We define the derivative message as follows:

$$\partial_X \lambda = \lambda$$
$$\partial_X \varepsilon = \varepsilon$$
$$\partial_X X = \varepsilon$$
$$\partial_X Y = Y, \quad X \neq Y$$
$$\partial \{X\} m = \partial_X m$$
$$\partial [\overline{X}] m = \partial_{\{\mathscr{X}_m \backslash X\}} m$$
$$\partial_X f(m) = f(\partial_X m), \quad f \in \Sigma$$
$$\partial_{S_1 \cup S_2} m = \partial_{S_1} \partial_{S_2} m$$

Then, we apply a safe function F to derivative messages. For an atom in the static neighborhood (i.e. in ∂m), we evaluate its security with no respect to variables. Else, for any message replacing a variable, it is evaluated as a constant block, whatever its content, and with no respect to other variables, if any. This is described by Definition 6.

Definition 6 Let $m \in \mathcal{M}_p^{\mathcal{G}}$ and $m\sigma$ be a valid trace. $\forall \lambda \in \mathcal{A}(m)$, $\forall \sigma \in \Gamma$, we denote by:

$$F(\lambda, \partial[\bar{\lambda}]m) = F(\lambda\sigma, \partial[\lambda\sigma]m\sigma) = \begin{cases} F(\lambda, \partial m) & \forall \lambda \in \mathcal{A}(\partial m), \\ F(\lambda, \partial_{\{\mathcal{X}_m \setminus X\}}m) & \forall \lambda \in \mathcal{X}_m. \end{cases}$$

The application in Definition 6 could not be used to analyze protocols. It is error-prone. Let us examine its deficiency in the Example 3.

Example 3 Let m_1 and m_2 be two messages of $\mathcal{M}_p^{\mathcal{G}}$ such that $m_1 = \{\lambda.D.X\}_{k_{ab}}$ and $m_2 = \{\lambda.Y\}_{k_{ab}}$ and $\ulcorner \lambda \urcorner = \{A, B\}$. Let $m = \{\lambda.D.B\}_{k_{ab}}$ be a close message in a valid trace.

$$F_{MAX}^{EK}(\lambda, \partial[\bar{\lambda}]m) = \begin{cases} \{A, B, D\}, & \text{if } m = m_1\sigma_1 | X\sigma_1 = B, \\ \{A, B\}, & \text{if } m = m_2\sigma_2 | Y\sigma_2 = D.B \end{cases}$$

Therefore, $F_{MAX}^{EK}(\lambda, \partial[\bar{\lambda}]m)$ is not a function on $m\sigma$ (i.e. it returns two possible values for the same preimage).

The witness-function in Definition 7 fixes this deficiency: it looks for all the origins m of the substituted message $m\sigma$ in the generalized roles, applies the application in Definition 6 and returns the minimum that obviously exists and is unique in the security lattice.

Definition 7 *(witness-function)* Let $m \in \mathcal{M}_p^{\mathcal{G}}$, $X \in \mathcal{X}_m$ and $m\sigma$ be a valid trace. Let p be a protocol and F be a $\mathcal{C}$-safe Function. We define a witness-function $\Phi_{p,F}$ for all $\lambda \in \mathcal{A}(m)$ and $\sigma \in \Gamma$, as follows:

$$\Phi_{p,F}(\lambda\sigma, m\sigma) = \bigsqcap_{\substack{(m',\sigma') \in \mathcal{M}_p^{\mathcal{G}} \times \Gamma \\ m'\sigma' = m\sigma}} F(\lambda, \partial[\bar{\lambda}]m'\sigma')$$

A witness-function $\Phi_{p,F}$ is safe when F is. Indeed, it is easy to verify that it is well-built. It is invariant-by-intruder as well since the returned values (principal identities) are those returned by F applied to derivative messages of the origins of $m\sigma$. Derivation does not add new candidates, it just removes some of them, but returns always candidates from the same invariant section by the intruder when the message is substituted.

Since the target of the witness-functions is to analyze protocols statically and since it still depends on σ (runs), we will bind it in two static bounds and use them for analysis instead of the witness-function itself. Lemma 1 provides these bounds.

Lemma 1 (witness-function Bounds) *Let $m \in \mathcal{M}_p^{\mathcal{G}}$. Let F be a $\mathcal{C}$-safe function and $\Phi_{p,F}$ be a witness-function. For all $\sigma \in \Gamma$ we have:*

$$F(\lambda, \partial[\bar{\lambda}]m) \sqsupseteq \Phi_{p,F}(\lambda\sigma, m\sigma) \sqsupseteq \bigsqcap_{\substack{(m'\sigma') \in \mathcal{M}_p^{\mathcal{G}} \times \Gamma \\ m'\sigma' = m\sigma'}} F(\lambda, \partial[\bar{\lambda}]m'\sigma')$$

For a secret $\lambda\sigma$ in a substituted message $m\sigma$, the upper-bound $F(\lambda, \partial[\bar{\lambda}]m)$ evaluates its security from one confirmed origin m in the generalized roles, the witness-function $\Phi_{p,F}(\lambda, m\sigma)$ from the set of the exact origins of $m\sigma$ (when running). The message m is obviously one of them. The lower-bound $\sqcap_{\substack{(m'\sigma')\in\mathcal{M}_p^{\mathcal{G}}\times\Gamma \\ m'\sigma'=m\sigma'}} F(\lambda, \partial[\bar{\lambda}]m'\sigma')$ evaluates it from the set of all the messages that are unifiable with m. This set naturally includes the set of definition of the witness-function since unifications include substitutions. Unifications in the lower-bound trap any intrusion (odd principal identities). Please notice that both the upper-bound and the lower-bound are static (independent of σ).

Theorem 2 (Analysis Theorem) *Let p be a protocol. Le F be a safe function. Let $\Phi_{p,F}$ be a witness-function. p keeps its secrect inputs if:*
$$\forall R.r \in R_G(p), \forall \lambda \in A(r^+) \text{ we have:}$$

$$\sqcap_{\substack{(m'\sigma')\in\mathcal{M}_p^{\mathcal{G}}\times\Gamma \\ m'\sigma'=r^+\sigma'}} F(\lambda, \partial[\bar{\lambda}]m'\sigma') \sqsupseteq \ulcorner\lambda\urcorner \sqcap F(\lambda, \partial[\bar{\lambda}]R^-)$$

This theorem states a static criterion for secrecy. It derives directly from Theorem 1 and Lemma 1. This allows us to analyze a protocol from within its generalized roles (finite set) and send any decision made-on to valid traces.

5 Analysis of the Woo-Lam Protocol (Amended Version) with a Witness-Function

The original version of the Woo-Lam protocol has been proven incorrect for secrecy by too many means [11–13]. This is due to a well-known flaw that it involves and described in Fig. 1. Hereafter, we analyze an amended version of this protocol with a witness-function and we prove that it is correct for secrecy. This version is denoted by p' in Table 2.

The role-based specification of p is $\mathcal{R}_G(p') = \{\mathcal{A}_G^1, \mathcal{A}_G^2, \mathcal{B}_G^1, \mathcal{B}_G^2, \mathcal{B}_G^3, \mathcal{S}_G^1\}$, where the generalized roles $\mathcal{A}_G^1, \mathcal{A}_G^2$ of A are as follows:

$$\mathcal{A}_G^1 = \langle i.1, A \rightarrow I(B) : A \rangle$$
$$\mathcal{A}_G^2 = \langle i.1, A \rightarrow I(B) : A \rangle.$$
$$\langle i.2, I(B) \rightarrow A : X \rangle.$$
$$\langle i.3, A \rightarrow I(B) : \{B.k_{ab}^i\}_{k_{as}} \rangle$$

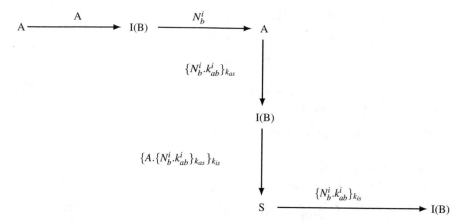

Fig. 1 Flaw in the Woo-Lam protocol (original version)

Table 2 The Woo-Lam Protocol (amended version)

$$
\begin{aligned}
p' := \ &\langle 1, A \to B : A \rangle . \\
&\langle 2, B \to A : N_b \rangle . \\
&\langle 3, A \to B : \{B.k_{ab}\}_{k_{as}} \rangle . \\
&\langle 4, B \to S : \{A.N_b.\{B.k_{ab}\}_{k_{as}}\}_{k_{bs}} \rangle . \\
&\langle 5, S \to B : \{N_b.\{A.k_{ab}\}_{k_{bs}}\}_{k_{bs}} \rangle
\end{aligned}
$$

The generalized roles $\mathscr{B}_G^1$, $\mathscr{B}_G^2$, $\mathscr{B}_G^3$ of B are as follows:

$$
\begin{aligned}
\mathscr{B}_G^1 \ = \ &\langle i.1, I(A) \ \to \ B \quad : \ A \rangle . \\
&\langle i.2, B \ \to \ I(A) \ : \ N_b^i \rangle \\
\mathscr{B}_G^2 \ = \ &\langle i.1, I(A) \ \to \ B \quad : \ A \rangle . \\
&\langle i.2, B \ \to \ I(A) \ : \ N_b^i \rangle . \\
&\langle i.3, I(A) \ \to \ B \quad : \ Y \rangle . \\
&\langle i.4, B \ \to \ I(S) \ : \ \{A.N_b^i.Y\}_{k_{bs}} \rangle \\
\mathscr{B}_G^3 \ = \ &\langle i.1, I(A) \ \to \ B \quad : \ A \rangle . \\
&\langle i.2, B \ \to \ I(A) \ : \ N_b^i \rangle . \\
&\langle i.3, I(A) \ \to \ B \quad : \ Y \rangle . \\
&\langle i.4, B \ \to \ I(S) \ : \ \{A.N_b^i.Y\}_{k_{bs}} \rangle . \\
&\langle i.5, I(S) \ \to \ B \quad : \ \{N_b^i.\{A.Z\}_{k_{bs}}\}_{k_{bs}} \rangle
\end{aligned}
$$

The generalized role $\mathscr{S}_G^1$ of S is as follows:

$$
\begin{aligned}
\mathscr{S}_G^1 \ = \ &\langle i.4, I(B) \ \to \ S \quad : \ \{A.U.\{B.V\}_{k_{as}}\}_{k_{bs}} \rangle . \\
&\langle i.5, S \ \to \ I(B) \ : \ \{U.\{A.V\}_{k_{bs}}\}_{k_{bs}} \rangle
\end{aligned}
$$

Let us have a context of verification such that:

$$\ulcorner k_{as}\urcorner = \{A, S\}; \quad \ulcorner k_{bs}\urcorner = \{B, S\}; \quad \ulcorner k_{ab}^i\urcorner = \{A, B, S\};$$
$$\ulcorner N_b^i\urcorner = \bot; \quad \forall A \in \mathscr{I}, \ulcorner A\urcorner = \bot.$$

The principal identities are not analyzed since they are set public in the context.
Let $F = F_{MAX}^{EK}$; $\Phi_{p',F} = \Phi_{p',F_{MAX}^{EK}}$;

We denote by $\Phi'_{p',F}(\lambda, m)$ the lower-bound $\displaystyle\prod_{\substack{(m'\sigma') \in \mathscr{M}_p^{\mathscr{G}} \times \Gamma \\ m'\sigma' = m\sigma'}} F(\lambda, \partial[\bar{\lambda}]m'\sigma')$ of the

witness-function $\Phi_{p',F}(\lambda, m)$.

The set of messages generated by p' is $\mathscr{M}_{p'}^{\mathscr{G}} = \{A_1, X_1, \{B_1.K_{A_2B_1}^i\}_{K_{A_2S_1}},$
$A_3, N_{B_2}^i, Y_1, \{A_4.N_{B_3}^i.Y_2\}_{K_{B_3S_2}}, \{N_{B_4}^i.\{A_5.Z_1\}_{K_{B_4S_3}}\}_{K_{B_4S_3}}, \{A_6.U_1.\{B_5.V_1\}_{K_{A_6S_4}}\}_{K_{B_5S_4}},$
$\{U_2.\{A_7.V_2\}_{K_{B_6S_5}}\}_{K_{B_6S_5}}\}.$

After elimination of duplicates, $\mathscr{M}_{p'}^{\mathscr{G}} = \{A_1, X_1, \{B_1.K_{A_2B_1}^i\}_{K_{A_2S_1}}, N_{B_2}^i,$
$\{A_4.N_{B_3}^i.Y_2\}_{K_{B_3S_2}}, \{N_{B_4}^i.\{A_5.Z_1\}_{K_{B_4S_3}}\}_{K_{B_4S_3}}, \{A_6.U_1.\{B_5.V_1\}_{K_{A_6S_4}}\}_{K_{B_5S_4}}, \{U_2.\{A_7.$
$V_2\}_{K_{B_6S_5}}\}_{K_{B_6S_5}}\}.$

The variables are denoted by $X_1, Y_2, Z_1, U_1, U_2, V_1$ and V_2.

5.1 Analysis of the Generalized Roles of A

As defined in the generalized role A, an agent A can participate in some session S^i in which she receives an unkown message X and sends the message $\{B.k_{ab}^i\}_{k_{as}}$. This is described by the following rule:

$$S^i : \frac{X}{\{B.k_{ab}^i\}_{k_{as}}}$$

- Analysis of the messages exchanged in S^i:

1. For k_{ab}^i:

   (a) When receiving: $R_{S^i}^- = X$ (when receiving, we use the upper-bound

   $$F(k_{ab}^i, \partial[\overline{k_{ab}^i}]X) = F(k_{ab}^i, \varepsilon) = \mathrm{T} \tag{1}$$

   (b) When sending: $r_{S^i}^+ = \{B.k_{ab}^i\}_{k_{as}}$ (on sending, we use the lower-bound)

   $$\forall k_{ab}^i.\{(m', \sigma') \in \mathscr{M}_{p'}^{\mathscr{G} \times \Gamma} | m'\sigma' = r_{S^i}^+ \sigma'\}$$
   $$= \forall k_{ab}^i.\{(m'\sigma') \in \mathscr{M}_{p'}^{\mathscr{G} \times \Gamma} | m'\sigma' = \{B.k_{ab}^i\}_{k_{as}} \sigma'\}$$
   $$= \{(\{B_1.K_{A_2B_1}^i\}_{K_{A_2S_1}}, \sigma'_1)\}$$

such that:

$$\sigma'_1 = \{B_1 \mapsto B, K^i_{A_2B_1} \mapsto k^i_{ab}, K_{A_2S_1} \mapsto k_{as}\}$$

$$\Phi'_{p',F}(k^i_{ab}, \{B.k^i_{ab}\}_{k_{as}})$$

$$= \{\textit{Definition of the lower-bound of the witness-function}\}$$

$$F(k^i_{ab}, \partial\overline{[k^i_{ab}]}\{B_1.K^i_{A_2B_1}\}_{K_{A_2S_1}} \sigma'_1)$$

$$= \{\textit{Extracting the static neighborhood}\}$$

$$F(k^i_{ab}, \partial\overline{[k^i_{ab}]}\{B.k^i_{ab}\}_{k_{as}} \sigma'_1) = \{\textit{Definition 6}\}$$

$$F(k^i_{ab}, \partial\overline{[k^i_{ab}]}\{B.k^i_{ab}\}_{k_{as}}) = \{\textit{Derivation in Definition 5}\}$$

$$F(k^i_{ab}, \{B.k^i_{ab}\}_{k_{as}}) = \{\textit{Since } F = F^{EK}_{MAX}\}$$

$$\{B, A, S\}$$

Then, we have:

$$\Phi'_{p',F}(k^i_{ab}, \{B.k^i_{ab}\}_{k_{as}}) = \{B, A, S\} \tag{2}$$

2. Compliance with Theorem 2:
   From 1 and 2, we have:

$$\Phi'_{p',F}(k^i_{ab}, \{B.k^i_{ab}\}_{k_{as}}) \sqsupseteq \ulcorner k^i_{ab} \urcorner \sqcap F(k^i_{ab}, \partial\overline{[k^i_{ab}]}X) \tag{3}$$

From 3, we have: the messages exchanged in the session S^i (i.e. k^i_{ab}) respect Theorem 2. (I)

5.2 Analysis of the Generalized Roles of B

As defined in the generalized roles of B, an agent B can participate in two subsequent sessions: S^i and S^j such that $j > i$. In the former session S^i, the agent B receives the identity A and sends the nonce N^i_b. In the subsequent session S^j, she receives an unknown message Y and she sends the message $\{A.N^i_b.Y\}_{k_{bs}}$. This is described by the following rules:

$$S^i : \frac{A}{N^i_b} \qquad S^j : \frac{Y}{\{A.N^i_b.Y\}_{k_{bs}}}$$

– Analysis of the messages exchanged in S^i:

1. For N_b^i:
   Since N_b^i is declared public in the context (i.e. $\ulcorner N_b^i \urcorner = \bot$), then, we have directly:

$$\Phi'_{p',F}(N_b^i, N_b^i) \sqsupseteq \ulcorner N_b^i \urcorner \sqcap F(N_b^i, \partial[\overline{N_b^i}]A) = \bot \qquad (4)$$

– Analysis of the messages exchanged in S^j:

1. For N_b^i:
   Since N_b^i is declared public in the context (i.e. $\ulcorner N_b^i \urcorner = \bot$), then, we have directly:

$$\Phi'_{p',F}(N_b^i, \{A.N_b^i.Y\}_{k_{bs}}) \sqsupseteq \ulcorner N_b^i \urcorner \sqcap F(N_b^i, \partial[\overline{N_b^i}]Y) = \bot \qquad (5)$$

2. $\forall Y$:
   Since when receiving, we have $F(Y, \partial[\overline{Y}]Y) = F(Y, Y) = \bot$, then, we have directly:

$$\Phi'_{p',F}(Y, \{A.N_b^i.Y\}_{k_{bs}}) \sqsupseteq \ulcorner Y \urcorner \sqcap F(Y, \partial[\overline{Y}]Y) = \bot \qquad (6)$$

3. Compliance with Theorem 2:
   From 4, 5 and 6, we have: the messages exchanged in the sessions S^i and S^j respect Theorem 2. (II)

5.3   Analysis of the Generalized Roles of S

As defined in the generalized role S, an agent S can participate in some session S^i in which she receives the message $\{A.U.\{B.V\}_{k_{as}}\}_{k_{bs}}$ and sends the message $\{U.\{A.V\}_{k_{bs}}\}_{k_{bs}}$. This is described by the following rule:

$$S^i : \frac{\{A.U.\{B.V\}_{k_{as}}\}_{k_{bs}}}{\{U.\{A.V\}_{k_{bs}}\}_{k_{bs}}}$$

1. $\forall U$:

   (a) When receiving: $R_{S^i}^- = \{A.U.\{B.V\}_{k_{as}}\}_{k_{bs}}$ (when receiving, we use the upper-bound)

$$F(U, \partial[\overline{U}]\{A.U.\{B.V\}_{k_{as}}\}_{k_{bs}}) = F(U, \{A.U.\{B\}_{k_{as}}\}_{k_{bs}}) = \{A, B, S\} \qquad (7)$$

(b) When sending: $r_{S^i}^+ = \{U.\{A.V\}_{k_{bs}}\}_{k_{bs}}$ *(on sending, we use the lower-bound)*

$$\forall U.\{(m', \sigma') \in \mathscr{M}_{p'}^{\mathscr{G}} | m'\sigma' = r_{S^i}^+ \sigma'\}$$
$$= \forall U.\{(m', \sigma') \in \mathscr{M}_{p'}^{\mathscr{G}} | m'\sigma' = \{U.\{A.V\}_{k_{bs}}\}_{k_{bs}} \sigma'\}$$
$$= \{(\{\{U_2.\{A_7.V_2\}_{K_{B_6 S_5}}\}_{K_{B_6 S_5}}, \sigma'_1)\}$$

such that:

$$\sigma'_1 = \{U_2 \mapsto U, A_7 \mapsto A, V_2 \mapsto V, K_{B_6 S_5} \mapsto k_{bs}\}$$
$$\Phi'_{p',F}(U, \{U.\{A.V\}_{k_{bs}}\}_{k_{bs}})$$
$$= \{Definition \ of \ the \ lower\text{-}bound \ of \ the \ witness\text{-}function\}$$
$$F(U, \partial[\overline{U}]\{U_2.\{A_7.V_2\}_{K_{B_6 S_5}}\}_{K_{B_6 S_5}} \sigma'_1)$$
$$= \{Extracting \ the \ static \ neighborhood\}$$
$$F(U, \partial[\overline{U}]\{U_2.\{A.V\}_{k_{bs}}\}_{k_{bs}} \sigma'_1) = \{Definition \ 6\}$$
$$F(U_2, \partial[\overline{U_2}]\{U_2.\{A.V\}_{k_{bs}}\}_{k_{bs}}) = \{Derivation \ in \ Definition \ 5\}$$
$$F(U_2, \{U_2.\{A\}_{k_{bs}}\}_{k_{bs}}) = \{Since \ F = F_{MAX}^{EK}\}$$
$$\{A, B, S\}$$

Then, we have:

$$\Phi'_{p',F}(U, \{U.\{A.V\}_{k_{bs}}\}_{k_{bs}}) = \{A, B, S\} \tag{8}$$

2. $\forall V$:

(a) When receiving: $R_{S^i}^- = \{A.U.\{B.V\}_{k_{as}}\}_{k_{bs}}$ *(when receiving, we use the upper-bound)*

$$F(V, \partial[\overline{V}]\{A.U.\{B.V\}_{k_{as}}\}_{k_{bs}}) = F(V, \{A.\{B.V\}_{k_{as}}\}_{k_{bs}})$$

$$= \begin{cases} \{A, B, S\} & \begin{array}{l} if \ k_{as} \ is \ the \ external \ protective \ key \ of \ V \\ in \ the \ received \ message \{A.\{B.V\}_{k_{as}}\}_{k_{bs}} \end{array} \\ \{A, B, S\} & \begin{array}{l} if \ k_{bs} \ is \ the \ external \ protective \ key \ of \ V \\ in \ the \ received \ message \{A.\{B.V\}_{k_{as}}\}_{k_{bs}} \end{array} \end{cases}$$

$$= \{A, B, S\}$$

Then, we have:

$$F(V, \partial[\overline{V}]\{A.U.\{B.V\}_{k_{as}}\}_{k_{bs}}) = \{A, B, S\} \tag{9}$$

(b) When sending: $r_{S^i}^+ = \{U.\{A.V\}_{k_{bs}}\}_{k_{bs}}$ *(on sending, we use the lower-bound)*

$$\forall V.\{(m',\sigma') \in \mathscr{M}_{p'}^{\mathscr{G}} | m'\sigma' = r_{S^i}^+ \sigma'\}$$
$$= \forall V.\{(m',\sigma') \in \mathscr{M}_{p'}^{\mathscr{G}} | m'\sigma' = \{U.\{A.V\}_{k_{bs}}\}_{k_{bs}} \sigma'\}$$
$$= \{(\{\{U_2.\{A_7.V_2\}_{K_{B_6 S_5}}\}_{K_{B_6 S_5}}, \sigma_1'),$$
$$(\{N_{B_4}^i.\{A_5.Z_1\}_{K_{B_4 S_3}}\}_{K_{B_4 S_3}}, \sigma_2')\}$$

such that:

$$\begin{cases} \sigma_1' = \{U_2 \mapsto U, A_7 \mapsto A, V_2 \mapsto V, K_{B_6 S_5} \mapsto k_{bs}\} \\ \sigma_2' = \{U \mapsto N_{B_4}^i, A_5 \mapsto A, Z_1 \mapsto V, K_{B_4 S_3} \mapsto k_{bs}\} \end{cases}$$

$\Phi'_{p',F}(V, \{U.\{A.V\}_{k_{bs}}\}_{k_{bs}})$
   $= \{\text{\textit{Definition of the lower-bound of the witness-function}}\}$
$F(V, \partial\overline{[V]}\{U_2.\{A_7.V_2\}_{K_{B_6 S_5}}\}_{K_{B_6 S_5}} \sigma_1') \sqcap F(V, \partial\overline{[V]}\{N_{B_4}^i.\{A_5.Z_1\}_{K_{B_4 S_3}}\}_{K_{B_4 S_3}} \sigma_2')$
   $= \{\text{\textit{Extracting the static neighborhood}}\}$
$F(V, \partial\overline{[V]}\{U.\{A.V_2\}_{k_{bs}}\}_{k_{bs}} \sigma_1') \sqcap F(V, \partial\overline{[V]}\{N_{B_4}^i.\{A.V_2\}_{k_{bs}}\}_{k_{bs}} \sigma_2') = \{\text{\textit{Definition 6}}\}$
$F(V_2, \partial\overline{[V_2]}\{U.\{A.V_2\}_{k_{bs}}\}_{k_{bs}}) \sqcap F(V_2, \partial\overline{[V_2]}\{N_{B_4}^i.\{A.V\}_{k_{bs}}\}_{k_{bs}}) = \{\text{\textit{Derivation in Definition 5}}\}$
$F(V_2, \{\{A.V\}_{k_{bs}}\}_{k_{bs}}) \sqcap F(V_2, \{N_{B_4}^i.\{A.V\}_{k_{bs}}\}_{k_{bs}}) = \{\text{\textit{Since }} F = F_{MAX}^{EK}\}$
$\{A, B, S\}$

Then, we have:

$$\Phi'_{p',F}(V, \{U.\{A.V\}_{k_{bs}}\}_{k_{bs}}) = \{A, B, S\} \tag{10}$$

3. Compliance with Theorem 2:

   - $\forall U$, from 7 and 8, we have:

$$\Phi'_{p',F}(U, \{U.\{A.V\}_{k_{bs}}\}_{k_{bs}}) \sqsupseteq \ulcorner U \urcorner \sqcap F(U, \partial\overline{[U]}\{A.U.\{B.V\}_{k_{as}}\}_{k_{bs}}) \tag{11}$$

   - $\forall V$, from 9 and 10, we have:

$$\Phi'_{p',F}(V, \{U.\{A.V\}_{k_{bs}}\}_{k_{bs}}) \sqsupseteq \ulcorner V \urcorner \sqcap F(V, \partial\overline{[V]}\{A.U.\{B.V\}_{k_{as}}\}_{k_{bs}}) \tag{12}$$

From 11 to 12, we have: the messages exchanged in the session S^i respect Theorem 2 (III).

Table 3 Compliance of the Woo-Lam protocol (amended version) with Theorem 2

	λ	Role	R^-	r^+	Theorem 2
1	k_{ab}^i	A	X	$\{B.k_{ab}^i\}_{k_{as}}$	Ok
2	$\forall X$	A	X	$\{B.k_{ab}^i\}_{k_{as}}$	Ok
3	N_b^i	B	A	N_b^i	Ok
4	$\forall Y$	B	Y	$\{A.N_b^i.Y\}_{k_{bs}}$	Ok
5	N_b^i	B	Y	$\{A.N_b^i.Y\}_{k_{bs}}$	Ok
6	$\forall U$	S	$\{A.U.\{B.V\}_{k_{as}}\}_{k_{bs}}$	$\{A.V\}_{k_{bs}}$	Ok
7	$\forall V$	S	$\{A.U.\{B.V\}_{k_{as}}\}_{k_{bs}}$	$\{A.V\}_{k_{bs}}$	Ok

6 Results and Interpretation

The results of analysis of the amended version of the Woo-Lam protocol are summarized in Table 3. From Table 3, we conclude that this version fully respects Theorem 2. Hence, this protocol keeps its secrect inputs.

7 Related Works

Our witness-functions are comparable to the rank-functions of Schneider [14] and the interpretation-functions of Houmani [15–18]. Unlike the rank-functions, the witness-function are easy to build and easy to use. The rank-functions require CSP [19, 20] and are difficult to search in a protocol [11]. They could even not exist [21]. Unlike the interpretation-functions, the witness-functions do not dictate that a message must be protected by the direct key. Any further protective key could define a witness-function. Our functions do not depend on variables thanks to their static bounds. That is a major fact. All that makes our witness-function more flexible and would allow us to prove correctness of a wider range of protocols.

8 Conclusion and Future Work

In this work, we presented a new framework to analyze statically cryptographic protocols for secrecy using the witness-functions. We successfully tested them on an amended version of the Woo-Lam protocol. In a future work, we will test them on protocols with theories [22–24] and on compose protocols [25–27]. We believe that our witness-functions will help to treat these problems.

References

1. Fattahi, J., Mejri, M., Houmani, H.: Secrecy by witness-functions on increasing protocols. In: The 6th International Conference on Electronics, Computers and Artificial Intelligence (ECAI), Bucharest, Romania, pp. 1–6. IEEE, Oct 2014
2. Fattahi, J., Mejri, M., Houmani, H.: Secrecy by witness functions. In: 5th Proceedings of the Formal Methods for Security Workshop co-located with the PetriNets-2014 Conference, pp. 34–52, 2014
3. Fattahi, J., Mejri, M., Houmani, H.: New functions for secrecy on real protocols. In: Fourth International Conference on Computer Science, Engineering and Applications (ICCSEA 2014), Chennai, India, pp. 229–250, 2014
4. Fattahi, J., Mejri, M., Houmani, H.: A Semi-Decidable Procedure for Secrecy in Cryptographic Protocols. ArXiv e-prints, Aug 2014
5. Fattahi, J., Mejri, M., Houmani, H.: Introduction to the witness-functions for secrecy in cryptographic protocols (in press). In: The 2014 International Conference on Networks and Information, Nanjing, China, 2014
6. Fattahi, J., Mejri, M., Houmani, H.: Relaxed conditions for secrecy in a role-based specification. Int. J. Inf. Secur. **1**, 33–36 (2014)
7. Dolev, D., Andrew, C.-C.Y.: On the security of public key protocols. IEEE Trans. Inf. Theory **29**(2):198–207 (1983)
8. Debbabi, M., Legaré, Y., Mejri, M.: An environment for the specification and analysis of cryptoprotocols. In: ACSAC, pp. 321–332, 1998
9. Debbabi, M., Mejri, M., Tawbi, N., Yahmadi, I.: Formal automatic verification of authentication crytographic protocols. In: ICFEM, pp. 50–59, 1997
10. Debbabi, M., Mejri, M., Tawbi, N., Yahmadi, I.: From protocol specifications to flaws and attack scenarios: an automatic and formal algorithm. In: WETICE, pp. 256–262, 1997
11. Shaikh, S.A., Bush, V.J.: Analysing the woo-lam protocol using csp and rank functions. In: WOSIS, pp. 3–12, 2005
12. Armando, A., Basin, D., Boichut, Y., Chevalier, Y., Compagna, L., Cuellar, J., Drielsma, P.H., Heám, P.-C., Mantovani, J., Mödersheim, S., von Oheimb, D., Rusinowitch, M., Santiago, J., Turuani, M., Viganò, L., Vigneron, L.: The AVISPA tool for the automated validation of internet security protocols and applications. In: Etessami, K., Rajamani, S.K. (eds.) Proceedings of the 17th International Conference on Computer Aided Verification (CAV'05), vol. 3576 of LNCS. Springer, New York (2005). Available at http://www.avispa-project.org/publications.html
13. Blanchet, B.: An automatic security protocol verifier based on resolution theorem proving (invited tutorial). In: 20th International Conference on Automated Deduction (CADE-20), Tallinn, Estonia, July 2005
14. Schneider, S.: Verifying authentication protocols in csp. IEEE Trans. Softw. Eng. **24**(9), 741–758 (1998)
15. Houmani, H., Mejri, M.: Practical and universal interpretation functions for secrecy. In: SECRYPT, pp. 157–164, 2007
16. Houmani, H., Mejri, M.: Ensuring the correctness of cryptographic protocols with respect to secrecy. In: SECRYPT, pp. 184–189, 2008
17. Houmani, H., Mejri, M.: Formal analysis of set and nsl protocols using the interpretation functions-based method. J. Comp. Netw. Commun. (2012)
18. Houmani, H., Mejri, M., Fujita, H.: Secrecy of cryptographic protocols under equational theory. Knowl. Based Syst. **22**(3), 160–173 (2009)
19. Schneider, S.: Security properties and csp. In: IEEE Symposium on Security and Privacy, pp. 174–187, 1996
20. Schneider, S.A., Delicata, R.: Verifying security protocols: an application of csp. In: 25 years Communicating Sequential Processes, pp. 243–263, 2004

21. Heather, J., Schneider, S.: A decision procedure for the existence of a rank function. J. Comput. Secur. **13**(2), 317–344 (2005)
22. Comon-Lundh, H., Cortier, V., Zalinescu, E.: Deciding security properties for cryptographic protocols. Application to key cycles. ACM Trans. Comput. Log. **11**(2) (2010)
23. Cortier, V., Delaune, S.: Decidability and combination results for two notions of knowledge in security protocols. J. Autom. Reason. **48**(4), 441–487 (2012)
24. Cortier, V., Kremer, S., Warinschi, B.: A survey of symbolic methods in computational analysis of cryptographic systems. J. Autom. Reason. **46**(3–4), 225–259 (2011)
25. Ciobaca, S., Cortier, V.: Protocol composition for arbitrary primitives. In: 2012 IEEE 25th Computer Security Foundations Symposium, vol. 0, pp. 322–336, 2010
26. Cortier, V.: Secure composition of protocols. In: TOSCA, pp. 29–32, 2011
27. Cortier, V., Delaune, S.: Safely composing security protocols. Formal Meth. Syst. Des. **34**(1), 1–36 (2009)

Quantification of Centralized/Distributed Secrecy in Stochastic Discrete Event Systems

Mariam Ibrahim, Jun Chen and Ratnesh Kumar

Abstract Unlike information, behaviors cannot be encrypted and may instead be protected by providing covers that generate indistinguishable observations from behaviors needed to be kept secret. Such a scheme may still leak information about secrets due to statistical difference between the occurrence probabilities of the secrets and their covers. Jensen-Shannon Divergence (JSD) is a possible means of quantifying statistical difference between two distributions and can be used to measure such information leak as is presented in this chapter. Using JSD, we quantify loss of secrecy in stochastic partially-observed discrete event systems in two settings: (i) the centralized setting, corresponding to a single attacker/observer, and (ii) the distributed collusive setting, corresponding to multiple attackers/observers, exchanging their observed information. In the centralized case, an observer structure is formed and used to aide the computation of JSD, in the limit, as the length of observations approach infinity to quantify the worst case loss of secrecy. In the distributed collusive case, channel models are introduced to extend the system model to capture the effect of exchange of observations, that allows the JSD computation of the centralized case to be applied over the extended model to measure the distributed secrecy loss.

M. Ibrahim (✉) · R. Kumar
Department of Electrical and Computer Engineering, Iowa State University,
Ames, IA 50011, USA
e-mail: mariami@iastate.edu

R. Kumar
e-mail: rkumar@iastate.edu

M. Ibrahim
Department of Mechatronics Engineering, German Jordanian University,
11180 Amman, Jordan

J. Chen
Idaho National Laboratory, Idaho Falls, USA
e-mail: junchen@iastate.edu

© Springer International Publishing Switzerland 2016
E. Pricop and G. Stamatescu (eds.), *Recent Advances in Systems Safety and Security*,
Studies in Systems, Decision and Control 62, DOI 10.1007/978-3-319-32525-5_2

1 Introduction

Growing progress in information and communication technologies has led to growth in eavesdropping and tampering of private communication or behaviors. In contrast to information, behaviors cannot be encrypted, and their *secrecy* can instead be attained through introduction of *covers* that ambiguate secrets in presence of partial observation. Many techniques for hiding secrets based on ambiguation schemes have been proposed as, *Steganography and Watermarking* [5, 16], *Network level Anonymization* [18], and *Software Obfuscation* [9].

Also, various notions of information secrecy have also been explored in literature. For example [1, 8, 21], examine non-interference, requiring that secrets (private variables) do not interfere with or influence the observables (public variables). Non-interference is a logical notion that can only indicate the presence or the absence of interference, but is unable to quantify the level of interference. In contrast, for stochastic systems, the mutual information between the private and public variables can be used to quantify the level of interference, and hence loss of secrecy [21]. Mutual information is only an average case measure, and a worst case measure can also be defined, using for example *min-entropy* [8]. Extension of the notion of non-interference over behaviors (sequences) was explored in [22], requiring that every secret behavior must be masked by a cover behavior so secrets do not uniquely influence the observations.

For probabilistic systems, mutual information can again be used to quantify the level of secrecy loss, and as shown in [3, 12], it can be related to a certain Jensen-Shannon Divergence (JSD) computation, which was first employed in [2] to measure the disparity between the distributions of a secret versus its cover as a way to quantify the secrecy. An approximation algorithm for computing an upper bound of JSD was also provided in [2]. In a similar spirit, Saboori and Hadjicostis [20] considered mutual information between the secret states and the observed behaviors, and required it to be upper bounded. Checking this is undecidable, and Saboori and Hadjicostis [19] proposed a stronger notion, requiring the probability of revealing secrets to remain upper bounded at each time step. In contrast, S_τ-secrecy [11] bounds the probability of revealing secrets over the set of *all* behaviors, as opposed to for each step. S_τ-secrecy can be viewed as a variant of the divergence used in [2]. More related works on secrecy can be found in a recent survey [13].

In this work, we employ the JSD based measure of secrecy loss, and propose a method to compute it for stochastic partially-observed discrete event systems (PODES), under two settings, centralized and distributed. In the centralized setting, the computation of "limiting" JSD measure, quantifying the worst case statistical difference that is defined over arbitrary long observation sequences, is presented. The proposed JSD based quantification for secrecy loss is shown to be equivalent to the mutual information between the distribution over the observations and that over the possible status of system execution (whether secret or cover) [3]. In the distributed collusive setting, there exist multiple observers/attackers that have their own personal observations, and also collude by exchanging their observations over channels, that introduce delays that are bounded. To compute JSD measure in this setting, we introduce channel models and use those to extend the system model as in [17], capturing own observations as well as the delayed communicated observations. The JSD computation approach of the centralized setting is then employed to the extended model to yield the JSD measure of the distributed collusive setting. Illustrative examples, including one concerning AES (Advanced Encryption Standard), are provided to demonstrate the proposed secrecy loss computation approaches.

2 Notation and Preliminaries

For an event set Σ, define $\overline{\Sigma} := \Sigma \cup \{\varepsilon\}$, where ε denotes "no-event". The set of all finite length event sequences over Σ, including ε is denoted as Σ^*, $\Sigma^+ := \Sigma^* - \{\varepsilon\}$, and Σ^n is the set of event sequences of length $n \in N$. A *trace* is a member of Σ^* and a *language* is a subset of Σ^*. We use $s \leq t$ to denote if $s \in \Sigma^*$ is a prefix of $t \in \Sigma^*$, and $|s|$ to denote the length of s or the number of events in s. For $L \subseteq \Sigma^*$, its prefix-closure is defined as $pr(L) := \{s \in \Sigma^* | \exists t \in \Sigma^* : st \in L\}$ and L is said to be prefix-closed (or simply closed) if $pr(L) = L$, i.e., whenever L contains a trace, it also contains all the prefixes of that trace. For $s \in \Sigma^*$ and $L \subseteq \Sigma^*$, $L \setminus s := \{t \in \Sigma^* | st \in L\}$ denotes the set of traces in L *after* s.

Stochastic PODES. We can model a stochastic PODES by a *stochastic automaton* $G = (X, \Sigma, \alpha, x_0)$, where X is the set of states, Σ is the finite set of events, $x_0 \in X$ is the initial state, and $\alpha : X \times \Sigma \times X \to [0, 1]$ is the probability transition function [10], and $\forall x \in X, \sum_{\sigma \in \Sigma} \sum_{x' \in X} \alpha(x, \sigma, x') = 1$. A non-stochastic PODES can be modeled as the same 4-tuple, but by replacing the transition function with $\alpha : X \times \Sigma \times X \to \{0, 1\}$, and a non-stochastic DES is deterministic if $\forall x \in X, \sigma \in \Sigma, \sum_{x' \in X} \alpha(x, \sigma, x') \in \{0, 1\}$. The transition probability function α can be generalized to $\alpha : X \times \Sigma^* \times X$ in a natural way: $\forall x_i, x_j \in X, s \in \Sigma^*, \sigma \in \Sigma$, $\alpha(x_i, s\sigma, x_j) = \sum_{x_k \in X} \alpha(x_i, s, x_k) \alpha(x_k, \sigma, x_j)$, and $\alpha(x_i, \varepsilon, x_j) = 1$ if $x_i = x_j$ and 0 otherwise.

Define the language generated by G as $L(G) := \{s \in \Sigma^* | \exists x \in X, \alpha(x_0, s, x) > 0\}$. For a given G, a *component* $C = (X_C, \alpha_C)$ of G is a "subgraph" of G, i.e., $X_C \subseteq X$ and $\forall x, x' \in X_C$ and $\sigma \in \Sigma$, $\alpha_C(x, \sigma, x') = \alpha(x, \sigma, x')$ whenever the latter is positive, and $\alpha_C(x, \sigma, x') = 0$ otherwise. C is said to be a *strongly connected component* (SCC) or *irreducible* if $\forall x, x' \in X_C, \exists s \in \Sigma^*$ such that $\alpha_C(x, s, x') > 0$. A SCC C is said to be *closed* if for each $x \in X_C, \sum_{\sigma \in \Sigma} \sum_{x' \in X_C} \alpha_C(x, \sigma, x') = 1$. The states which belong to a closed SCC are *recurrent states* and the remaining states (that do not belong to any closed SCC) are *transient states*. Another way to identify recurrent versus transient states is to consider the steady-state state distribution π^* as the fixed-point of $\pi^* = \pi^* \Omega$, where π^* is a row-vector with the same size as X, and Ω is the transition matrix with i jth entry being the transition probability $\sum_{\sigma \in \Sigma} \alpha(i, \sigma, j)$. (In case Ω is periodic with period $d \neq 1$, we consider the set of fixed-points of $\pi^* = \pi^* \Omega^d$.) Then any state i is recurrent if and only if there exists a reachable fixed point π^* such that the ith entry of π^* is nonzero. Identifying the set of recurrent states can be done polynomially, by the algorithm presented in [24].

Information Theoretic Notations. For a probability distribution p over discrete set A, its entropy is defined as $H(p) = -\sum_{a \in A} p(a) \log p(a)$. For two probability distributions p and q over A, their Kullback-Leibler (KL) divergences denoted as $D_{KL}(p,q)$, is defined as $D_{KL}(p,q) = \sum_{a \in A} p(a) \log \frac{p(a)}{q(a)}$. Given $\lambda_1 > 0$ and $\lambda_2 > 0$ satisfying $\lambda_1 + \lambda_2 = 1$, the Jensen-Shannon Divergence (JSD) between p and q under the weights (λ_1, λ_2), is defined as $D(p,q) = \lambda_1 D_{KL}(p, \lambda_1 p + \lambda_2 q) + \lambda_2 D_{KL}(q, \lambda_1 p + \lambda_2 q)$, which is equivalent to $D(p,q) = H(\lambda_1 p + \lambda_2 q) - \lambda_1 H(p) - \lambda_2 H(q)$ (for more details, refer to [6]). For two probability distributions p over A and q over B, their mutual information is defined as $I(p,q) = \sum_{a \in A, b \in B} Pr(a,b) \log \frac{Pr(a,b)}{p(a)q(b)}$, which can also be equivalently defined as $I(p,q) = H(p) - H(p|q)$, where the conditional entropy $H(p|q)$ is given as $H(p|q) = -\sum_{a \in A} p(a) \sum_{b \in B} Pr(b|a) \log Pr(b|a)$.

3 Illustrative Example: AES Side-Channel Attack

We consider a version of cache side-channel attack that can be used to compromise AES (Advanced Encryption Standard), adopted from [25]. The difference in access times of cache hit versus miss may be used to learn the AES key as described below.

AES is a symmetric crypto-system, which processes data blocks of 16, 24, or 32 bytes, using encryption keys of the same size as data, corresponding to "AES-16", "AES-24", or "AES-32". In what follows below, we consider AES-16 for illustration purposes. For encryption, the plain-text block is converted into the cipher-text block, both viewed as 4×4 array of bytes, in several rounds. The intermediate results of rounds are also of same sizes, and are termed "states". (For AES-16, the number of rounds Nr equals 10 [7, 15].) For setting up the keys for the various rounds, a key expansion algorithm is applied to an initial key $K^{(0)}$, outputting a linear array of 4-byte words, of length $4Nr$, corresponding to the keys $\{K^{(r)}, r = 1, \ldots, Nr\}$ for the future rounds.

Starting from a 16-byte plain-text $P = (p_0, \ldots, p_{15})$, encryption proceeds by computing a 16-byte intermediate state $x^{(r)} = (x_0^{(r)}, \ldots, x_{15}^{(r)})$ at each round r. The initial state $x^{(0)}$ is computed by $x_i^{(0)} = p_i \oplus k_i, i = 0, \ldots, 15$, and the next $Nr - 1$ rounds for $r = 0, \ldots, Nr - 2$ are computed as follows:

$$(x_0^{(r+1)}, x_1^{(r+1)}, x_2^{(r+1)}, x_3^{(r+1)}) \leftarrow T_0[x_0^{(r)}] \oplus T_1[x_5^{(r)}] \oplus T_2[x_{10}^{(r)}] \oplus T_3[x_{15}^{(r)}] \oplus K_0^{(r+1)}$$
$$(x_4^{(r+1)}, x_5^{(r+1)}, x_6^{(r+1)}, x_7^{(r+1)}) \leftarrow T_0[x_4^{(r)}] \oplus T_1[x_9^{(r)}] \oplus T_2[x_{14}^{(r)}] \oplus T_3[x_3^{(r)}] \oplus K_1^{(r+1)}$$
$$(x_8^{(r+1)}, x_9^{(r+1)}, x_{10}^{(r+1)}, x_{11}^{(r+1)}) \leftarrow T_0[x_8^{(r)}] \oplus T_1[x_{13}^{(r)}] \oplus T_2[x_2^{(r)}] \oplus T_3[x_7^{(r)}] \oplus K_2^{(r+1)}$$
$$(x_{12}^{(r+1)}, x_{13}^{(r+1)}, x_{14}^{(r+1)}, x_{15}^{(r+1)}) \leftarrow T_0[x_{12}^{(r)}] \oplus T_1[x_1^{(r)}] \oplus T_2[x_6^{(r)}] \oplus T_3[x_{11}^{(r)}] \oplus K_3^{(r+1)},$$

$$(1)$$

where the notation $T_n[m]$ denotes the index-m entry of table T_n that is used to store pre-computed transformations of states, involving the operations of substitute-bytes, shift-rows, and mix-columns. The last round is also computed using (1), except that tables $T_0, \ldots, T_3$ are replaced by tables $T_0^{(10)}, \ldots, T_3^{(10)}$, respectively. (The last round does not need the mix-columns operation and so uses different tables.)

An attacker may populate a cache line with an initial state $x_i = p_i \oplus k_i$, generated using a known plain-text p_i and a known key k_i, $i = 0, \ldots, 15$. When the host populates the same cache line with another initial state $x_i' = p_i' \oplus k_i'$, using another plain-text p_i', also known to the attacker, and a key k_i' that is unknown to the attacker, a cache hit, as indicated by a shorter access time, can indicate $x_i = x_i'$, implying $p_i \oplus k_i = p_i' \oplus k_i'$, from which the attacker can infer the unknown key, $k_i' = p_i' \oplus p_i \oplus k_i$. Thus each cache hit, which may be thought of host's cache line interfering with the attacker's cache line, provides an opportunity for an attacker to infer one byte of the key used by a host.

To provide additional protection against this vulnerability, the system may introduce random evictions of the cache. Figure 1a, b show the abstracted versions

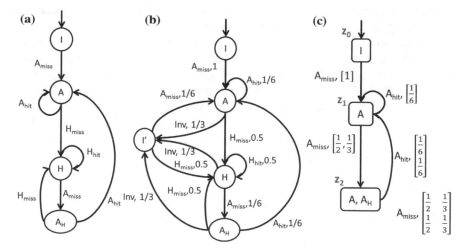

Fig. 1 **a** Cache side-channel attack model with no evictions, **b** cache side-channel attack model with random evictions, **c** observer for the cache side-channel attack with random evictions (reproduced from [3])

of the two cache architectures, with no protection and with added protection, respectively, where the models track the status of an individual cache line. (Similar models track other cache lines.) The 4 states in Fig. 1a are: "A" (occupied by the attacker and of low confidentiality), "H", "A_H" (occupied, respectively, by the host, and the attacker while occupied by the host in the previous step, both of high confidentiality), or "I" (invalid—that has no valid contents from attacker or host, and also of low confidentiality). If the host holds its own data in cache, its cache access results in a hit (H_{hit}), but if the attacker evicts the host's data in the cache lines by requesting cache access, it results in a miss (H_{miss}). The attackers cache hit and miss, A_{hit} and A_{miss} are dually defined. Note that the occurrence of A_{miss} or A_{hit} can be used to infer "H" or "A_H" states, using which one byte of the encryption key can be compromised. However, an attacker can only observe its own cache hits and misses (i.e., A_{hit} and A_{miss} are the only observable events). In Fig. 1b, random cache eviction is introduced by the system to invalidate the data, denoted by "Inv" event. This introduces ambiguity in the attacker's knowledge about the occupancy of the cache, i.e., when it observes a cache miss, it does not know whether it is due to the processor's eviction or due to the host's cache access. Then, in Fig. 1b, we can view $\{H, A_H\}$ to be the high confidential or "secret" states whereas $\{I, A, I'\}$ to be the low confidential or "cover" states, which present ambiguity against the "secret" states.

4 Quantification of Secrecy Loss in Centralized Setting

In this section, we study secrecy quantification in stochastic PODESs in the presence of a single attacker/observer, having partial observability of system behaviors for revealing sensitive system behaviors, as introduced below.

Secret/non-Secret Behaviors and Refined System. Certain system behaviors may be considered sensitive and hence secret, whereas the remaining behaviors act as covers for the secrets. Letting $L = L(G)$ denote the set of all behaviors (traces) of a stochastic PODES G as introduced in the notation section, suppose $K \subset L$ models the secret behaviors (also called a specification), while the remaining traces in $L - K$ act as its cover. K may be modeled by a deterministic acceptor $R = (Y, \Sigma, \beta, y_0)$ such that $L(R) = K$. By introducing a dump state D in R, and completing its transition function, we can obtain $\overline{R} = (\overline{Y}, \Sigma, \overline{\beta}, y_0)$, where $\overline{Y} = Y \cup D$, and $\forall \overline{y}, \overline{y}' \in \overline{Y}, \sigma \in \Sigma$,

$$\overline{\beta}(\overline{y}, \sigma, \overline{y}') := \begin{cases} \beta(\overline{y}, \sigma, \overline{y}') & \text{if } (\overline{y}, \overline{y}' \in Y) \wedge (\beta(\overline{y}, \sigma, \overline{y}') > 0), \\ 1 & \text{if } [(\overline{y} = \overline{y}' = D) \vee (\overline{y}' = D \wedge \sum_{y \in Y} \beta(\overline{y}, \sigma, y) = 0)]. \end{cases}$$

Then, the system model can be refined with respect to the specification to identify the secret and cover behaviors as *states* in the refined system $G^R = G \| \overline{R}$, and is given by $G^R = (X \times \overline{Y}, \Sigma, \gamma, (x_0, y_0))$, where $\forall (x, \overline{y}), (x', \overline{y}') \in X \times \overline{Y}, \sigma \in \Sigma$,

$$\gamma((x,\bar{y}),\sigma,(x',\bar{y'})) := \begin{cases} \alpha(x,\sigma,x') & \text{if}[(\bar{y},\bar{y'} \in Y \wedge \beta(\bar{y},\sigma,\bar{y'}) > 0) \vee (\bar{y} = \bar{y'} = D) \\ & \qquad \vee (\bar{y'} = D \wedge \sum_{y \in Y} \beta(\bar{y},\sigma,y) = 0)], \\ 0 & \text{otherwise.} \end{cases}$$

The events in Σ executed by the system are observed by an observer (an attacker or an adversary) through an observation mask $M : \overline{\Sigma} \rightarrow \overline{\Delta}$, where Δ is the set of observed symbols, and $M(\varepsilon) = \varepsilon$. ($M$ can be extended to Σ^* as follows: $M(\varepsilon) = \varepsilon$ and $\forall s \in \Sigma^*, \sigma \in \overline{\Sigma}, M(s\sigma) = M(s)M(\sigma)$.) The appendix describes the computation of an observer transition structure for G^R that can be used to track its evolution over its observed symbols Δ, and also the associated transition matrices $\{\Theta(\delta)|\delta \in \Delta\}$.

Jensen-Shannon Divergence Based Secrecy Quantification. The statistical difference between the conditional distributions of secrets versus covers over the system observations of a common length, provides a measure of the amount of secrecy leaked by a system. A possible way of measuring difference between two distributions is the JSD (Jensen Shannon Divergence) measure. Here we present a way to compute the JSD measure for stochastic PODESs. The JSD computation can be carried out over the refined system model following the method introduced in [3, 12], which we summarize here.

Given a length-n observation $o \in \Delta^n$, let $p_n(o)$ denote its probability. Then, since the occurrences of observations of length n are mutually disjoint, $\sum_{o \in \Delta^n} p_n(o) = 1$, i.e., p_n is a probability distribution over Δ^n. Then we can write its entropy as:

$$H(p_n) = -\sum_{o \in \Delta^n} p_n(o) \log p_n(o) = H(p_{n-1}) - \sum_{o \in \Delta^{n-1}} p_{n-1}(o) \sum_{\delta \in \Delta} p(\delta|o) \log p(\delta|o).$$

Observations in Δ^n can be generated by secrets (behaviors in K) or by covers (behaviors in $L - K$), and so we define two more probability distributions over Δ^n: probability that an observation $o \in \Delta^n$ is generated by some secret in K, denoted $p_n^s(o)$, versus that is generated by some cover in $L - K$, denoted $p_n^c(o)$:

$$p_n^s(o) := \frac{Pr(s \in K \cap M^{-1}(o))}{Pr(s \in K \cap M^{-1}(\Delta^n))}, \quad p_n^c(o) := \frac{Pr(s \in (L - K) \cap M^{-1}(o))}{Pr(s \in (L - K) \cap M^{-1}(\Delta^n))}.$$

Further, define $\lambda_n^s := Pr(s \in K \cap M^{-1}(\Delta^n))$ to be the probability of secrets and $\lambda_n^c := Pr(s \in (L - K) \cap M^{-1}(\Delta^n))$ to be the probability of covers, respectively, generating length-n observation. Then, it is easy to show that $\lambda_n^c := Pr(s \in (L - K) \cap M^{-1}(\Delta^n))$ for all $n \in \mathbb{N}$.

The ability of an intruder to identify secret versus cover behaviors based on observations of length-n, depends on the disparity between the two distributions p_n^s versus p_n^c: If p_n^s and p_n^c are identical, i.e., with "zero disparity", there is no way to statistically tell apart secrets from covers, and in that case there is perfect secrecy.

However, when p_n^s and p_n^c are different, then one could characterize the ability of an intruder to discriminate secrets from covers, based on length-n observations, using the JSD between p_n^s and p_n^c under the weights $(\lambda_n^s, \lambda_n^c)$, denoted $D(p_n^s, p_n^c) = H(\lambda_n^s p_n^s + \lambda_n^c p_n^c) - \lambda_n^s H(p_n^s) - \lambda_n^c H(p_n^c)$.

The following theorem from [3] shows that the JSD measure is indeed a useful measure of information revealed, as it equals the mutual information between the observations p_n and the status (whether secret or cover) of system executions. This status can be captured by a bi-valued random variable Λ_n, defined for each $n \in \mathbb{N}$, such that $Pr(\Lambda_n = s) = \lambda_n^s$ and $Pr(\Lambda_n = c) = \lambda_n^c$.

Theorem 1 ([3]). *The JSD between p_n^s and p_n^c equals the mutual information between Λ_n and p_n, i.e.,*

$$D(p_n^s, p_n^c) = I(\Lambda_n, p_n).$$

An intruder is likely to discriminate more if he/she observes for a longer period, and accordingly, our goal is to evaluate the worst-case loss of secrecy as obtain in the limit: $\lim_{n \to \infty} D(p_n^s, p_n^c)$. This worst-case JSD provides an upper bound to the amount of information leaked about secrets.

In order to compute JSD, we need to first compute the state-distribution of the observer, following each observation. Each observation $o \in \Delta^*$ results in a conditional state distribution $\pi(o)$, which can be computed recursively as follows: for any $o \in \Delta^*, \delta \in \Delta$: $\pi(\varepsilon) = \pi_0$ and $\pi(o\delta) = \frac{\pi(o) \times \Theta(\delta)}{\|\pi(o) \times \Theta(\delta)\|}$ [4], where π_0 is the initial state distribution, whereas the computation of transition matrix $\Theta(\delta)$ is given in the appendix. Let Π denote the set of all such conditional state distributions, and for each $\pi \in \Pi$ and $n \in \mathbb{N}$, denote $P_n(\pi) = Pr(o \in \Delta^n : \pi(o) = \pi)$, which is the probability that the set of all observations of length-n, upon which the conditional state distribution is π. For a state distribution π, define the following notations:

$$\lambda^{s|\pi} := \sum_{\delta \in \Delta} \pi \Theta(\delta) \mathscr{I}^s, \quad \lambda^{c|\pi} := \sum_{\delta \in \Delta} \pi \Theta(\delta) \mathscr{I}^c$$

$$p^{s|\pi}(\delta) := \frac{\pi \Theta(\delta) \mathscr{I}^s}{\lambda^{s|\pi}}, \quad p^{c|\pi}(\delta) := \frac{\pi \Theta(\delta) \mathscr{I}^c}{\lambda^{c|\pi}},$$

where $\mathscr{I}^s$ and $\mathscr{I}^c$ denote indicator column vectors of same size as number of states, with binary entries to identify the secret versus cover states (states reached by traces in K vs. $L - K$). Then, as shown in Lemma 4 of [3],

$$D(p_n^s, p_n^c) = H(\{\lambda_n^s, \lambda_n^c\}) + \sum_{\pi \in \Pi} P_{n-1}(\pi) \left[-H(\{\lambda^{s|\pi}, \lambda^{c|\pi}\}) + D(p^{s|\pi}, p^{c|\pi}) \right]. \quad (2)$$

In the limit when $n \to \infty$, if the distribution $P_n(\cdot)$ over Π converges to $P^*(\cdot)$, then $\lim_{n \to \infty} D(p_n^s, p_n^c)$ exists. See for example [14] for a condition under which such a convergence is guaranteed.

For an observer *Obs*, the computation of $\lim_{n\to\infty} D(p_n^s, p_n^c)$ using (2), requires the computation of $\lim_{n\to\infty} P_{n-1}(\pi)$ which can be accomplished with the help of an observer introduced in [3, 12]. The observer tracks the possible system states following each observation, and also allows the computation of the corresponding state distribution. We let *Obs* be an observer automaton with state set $Z \subseteq 2^{X \times \overline{Y}}$, so that each node $z \in Z$ of the observer is a subset of the refined system states, i.e., $z \subseteq (X, \overline{Y})$, and we use $|z|$ to denote the number of system states in z. *Obs* is initialized at node $z_0 = \{(x_0, y_0)\}$, and there is a transition labeled with $\delta \in \Delta$ from node z to z' if and only if every element of z' is reachable from some elements of z along a trace that ends in the only observation δ, i.e., $z' = \{(x', \overline{y}') \in X \times \overline{Y} : \exists (x, \overline{y}) \in z, L_{G^R}((x, \overline{y}), \delta, (x', \overline{y}')) \neq \emptyset\}$. Associated with this transition is the transition probability matrix $\Theta_{z,\delta,z'}$ of size $|z|$ by $|z'|$ (a submatrix of $\Theta(\delta)$ matrix given in the appendix), whose *ijth* element is $\theta_{i,\delta,j}$, which is the transition probability from *ith* element $(x, \overline{y})$ of z to *jth* element $(x', \overline{y}')$ of z' while producing the observation δ, and equals $\alpha(L_{G^R}((x, \overline{y}), \delta, (x', \overline{y}')))$.

Example 1 Consider the system, specification and refinement models of Fig. 2a–c, respectively, where $M(u) = \varepsilon$, $M(a) = a$ and $M(b) = b$. Then, the corresponding observer *Obs* is given in Fig. 2d, where each state in observer is a subset of states of the refined-system G^R, and transitions are on observed events that are labeled by their occurrence transition probability matrices.

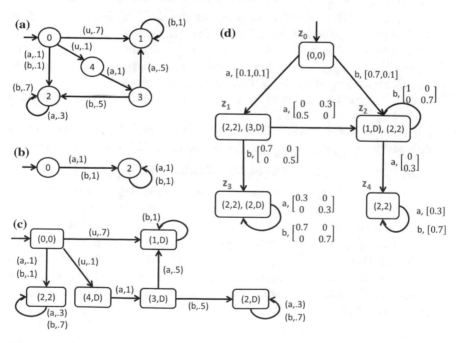

Fig. 2 **a** System model G, **b** specification for secrets, R, **c** refined system model G^R, **d** observer model (reproduced from [3])

Associated with each observation $o \in \Delta^*$, there is a reachable state distribution $\pi(o)$ as discussed earlier. Let the state z be reached in *Obs* following observation o. Then, obviously the number of positive elements of $\pi(o)$ is the same as the number of elements in z. Then, with a slight abuse of notation, we also use $\pi(o)$ to denote the row-vector containing only positive elements, and of same size as the number of elements in the node reached by o in *Obs*. Then, $\pi(o)$ can also be recursively computed as follows: for any $o \in \Delta^*, \delta \in \Delta$: $\pi(\varepsilon) = 1$ and $\pi(o\delta) = \frac{\pi(o) \times \Theta_{z_o,\delta,z_{o\delta}}}{||\pi(o) \times \Theta_{z_o,\delta,z_{o\delta}}||}$, where z_o and $z_{o\delta}$ are the nodes reached in *Obs* following o and $o\delta$ respectively. Then, it can be seen that along any cycle in *Obs*, the distribution upon completing the cycle is a function of the distribution upon entering the cycle, through a sequence of transition matrix-multiplications and their normalizations. In case of steady-state, those two distributions will be the same, namely, a fixed point of that function. The following assumption is made as in [3, 12].

Assumption 1 ([3, 12]) Assume that for any sufficiently long observations $o_1 \leq o_2$, if Obs reaches the same node following o_1 and o_2, then $\pi(o_1) = \pi(o_2)$.

Then as shown in [3, 12], the following procedure computes the worst-case loss of secrecy $\lim_{n \to \infty} D(p_n^s, p_n^c)$, under Assumption 1.

1. Construct a $(\sum_z |z|) \times (\sum_z |z|)$ square matrix $\tilde{\Theta}$, whose *ijth* block is the $|z_i| \times |z_j|$ matrix $\sum_\delta \Theta_{z_i,\delta,z_j}$. Compute the fix point distribution associated with $\tilde{\Theta}$ by solving $\pi^* = \pi^* \tilde{\Theta}$, where π^* is a row vector of size $\sum_z |z|$. For each $z_i \in Z$, let $p(z_i)$ be the summation of the *ith* block of π^*, then z_i is *recurrent* if $p(z_i) > 0$. Also note that for each $z \in Z$, exists a sufficiently large N such that $p(z) = \sum_{o \in \Delta^N :o \text{ reaches } z} p_N(o)$. In other words, $p(z)$ computes the probability of all sufficiently long observations that reach the observer state z.

2. Obtain λ^s as the summation of the elements of π^* corresponding to the secret states, i.e., $\lambda^s := \pi^* \mathscr{I}^s$, and $\lambda^c = 1 - \lambda^s$.

3. For a set of recurrent nodes $\{z_1, z_2, \ldots, z_n\}$ that form a SCC, define a set of distributions $\{\pi_{z_1}^*, \pi_{z_2}^*, \ldots, \pi_{z_n}^*\}$ to be a set of steady state distributions if $\forall i, j, \delta$, such that Θ_{z_i,δ,z_j} is defined, the following holds: $\pi_{z_j}^* = \frac{\pi_{z_i}^* \Theta_{z_i,\delta,z_j}}{||\pi_{z_i}^* \Theta_{z_i,\delta,z_j}||}$, i.e., $\pi_{z_i}^*$ represents a steady state conditional distribution following a single sufficiently long observation, that reaches z_i. Note that in this case, any other extension of o that also reaches z_i will induce the same conditional distribution $\pi_{z_i}^*$. There may exist multiple sets of steady state distributions for a given set of recurrent nodes, denoted say as $\{\{\pi_{z_1,k}^*, \ldots, \pi_{z_n,k}^*\}, k \in \mathbb{N}\}$. Then, if steady-state always exists, for any sufficiently long observation that reaches a recurrent node z, there exists $k \in \mathbb{N}$ such that $\pi(o) = \pi_{z,k}^*$. Denote $p(z, k) := Pr[\{o | o \text{ reaches } z \text{ and } \pi(o) = \pi_{z,k}^*\}]$.

4. Let $\mathscr{I}_{z'}^s$ and $\mathscr{I}_{z'}^c$ be indicator column vectors with binary entries of size $|z'|$ for identifying within z', the secret and cover states, respectively. For each steady state distribution $\pi_{z,k}^*$ of each recurrent node z, define:

$$\lambda^{s|\pi_{z,k}^*} := \sum_{\delta \in \Delta} \pi_{z,k}^* \Theta_{z,\delta,z'} \mathscr{I}_{z'}^s, \quad \lambda^{c|\pi_{z,k}^*} := \sum_{\delta \in \Delta} \pi_{z,k}^* \Theta_{z,\delta,z'} \mathscr{I}_{z'}^c$$

$$p^{s|\pi_{z,k}^*}(\delta) := \frac{\pi_{z,k}^* \Theta_{z,\delta,z'} \mathscr{I}_{z'}^s}{\lambda^{s|\pi_{z,k}^*}}, \quad p^{c|\pi_{z,k}^*}(\delta) := \frac{\pi_{z,k}^* \Theta_{z,\delta,z'} \mathscr{I}_{z'}^c}{\lambda^{c|\pi_{z,k}^*}}.$$

5. Then, applying (2), the JSD between p_n^s and p_n^c when $n \to \infty$ is given by:

$$\lim_{n \to \infty} D(p_n^s, p_n^c) = H(\{\lambda^s, \lambda^c\})$$

$$+ \sum_{z:z \text{ is recurrent}} \sum_{k \in \mathbb{N}} p(z,k) \left[-H(\{\lambda^{s|\pi_{z,k}^*}, \lambda^{c|\pi_{z,k}^*}\}) + D(p^{s|\pi_{z,k}^*}, p^{c|\pi_{z,k}^*}) \right].$$

$$(3)$$

(Note when the set of steady state distributions is unique, then in that case, $k = 1$ and we have: $p(z,k) = p(z)$ in (3) above.)

Example 2 We revisit Example 1. Then based on *Obs* of Fig. 2d, the following computation illustrates the steps of JSD computation.

1. $\sum_z |z| = 8$ and so $\tilde{\Theta}$ is a 8×8 matrix with entries:
$\tilde{\Theta}(1,2) = \tilde{\Theta}(1,3) = \tilde{\Theta}(1,5) = 0.1$, $\tilde{\Theta}(3,4) = \tilde{\Theta}(3,7) = 0.5$, $\tilde{\Theta}(1,4) = \tilde{\Theta}(2,6) = \tilde{\Theta}(5,5) = 0.7$, $\tilde{\Theta}(2,5) = \tilde{\Theta}(5,8) = 0.3$, $\tilde{\Theta}(4,4) = \tilde{\Theta}(6,6) = \tilde{\Theta}(7,7) = \tilde{\Theta}(8,8) = 1$, and zeros elsewhere. Then, $\pi^* = [0 \ \ 0 \ \ 0 \ \ 0.75 \ \ 0 \ \ 0.07 \ \ 0.05 \ \ 0.13]$. Therefore, $p(z_0) = p(z_1) = 0$, $p(z_2) = 0.75$, $p(z_3) = 0.12$ and $p(z_4) = 0.13$.
2. Here $\mathscr{I}^s = [1 \ \ 1 \ \ 0 \ \ 0 \ \ 1 \ \ 1 \ \ 0 \ \ 1]^T$, $\mathscr{I}^c = [0 \ \ 0 \ \ 1 \ \ 1 \ \ 0 \ \ 0 \ \ 1 \ \ 0]^T$. And so, $\lambda^s = 0.2$ and $\lambda^c = 0.8$.
3. Here z_2, z_3 and z_4 are recurrent nodes, and each of them forms a SCC. We have $\pi_{z_2}^* = [1 \ \ 0]$, $\pi_{z_4}^* = [1]$, and while there are multiple solutions to the equation set $\pi_{z_3}^* = \frac{\pi_{z_3}^* \Theta_{z_3,a,z_3}}{\pi_{z_3}^* \Theta_{z_3,a,z_3}}$ and $\pi_{z_3}^* = \frac{\pi_{z_3}^* \Theta_{z_3,b,z_3}}{\pi_{z_3}^* \Theta_{z_3,b,z_3}}$, only $\pi_{z_3}^* = [0.5833 \ \ 0.4167]$ is reachable. Thus, each set of recurrent nodes is a singleton set, and each with a unique fixed-point distribution. Therefore, for each recurrent node z, $p(z,k) = p(z)$.
4. Here $\mathscr{I}_{z_2}^s = [0 \ \ 1]^T$, $\mathscr{I}_{z_2}^c = [1 \ \ 0]^T$, $\mathscr{I}_{z_3}^s = [1 \ \ 0]^T$, $\mathscr{I}_{z_3}^c = [0 \ \ 1]^T$, $\mathscr{I}_{z_4}^s = [1]^T$ and $\mathscr{I}_{z_4}^c = [0]^T$. For z_2 and $\pi_{z_2}^*$, $\lambda^{s|\pi_{z_2}^*} = 0$, $\lambda^{c|\pi_{z_2}^*} = 1$, $p^{c|\pi_{z_2}^*}(b) = \frac{\pi_{z_2}^* \Theta_{z_2,b,z_2} \mathscr{I}_{z_2}^c}{\lambda^{c|\pi_{z_2}^*}} = 1$, $p^{s|\pi_{z_2}^*}(a) = p^{c|\pi_{z_2}^*}(a) = p^{s|\pi_{z_2}^*}(b) = 0$. For z_3 and $\pi_{z_3}^*$, $\lambda^{s|\pi_{z_3}^*} = 0.5833$, $\lambda^{c|\pi_{z_3}^*} = 0.4167$, $p^{s|\pi_{z_3}^*}(a) = \frac{\pi_{z_3}^* \Theta_{z_3,a,z_3} \mathscr{I}_{z_3}^s}{\lambda^{s|\pi_{z_3}^*}} = 0.3$, $p^{s|\pi_{z_3}^*}(b) = \frac{\pi_{z_3}^* \Theta_{z_3,b,z_3} \mathscr{I}_{z_3}^s}{\lambda^{s|\pi_{z_3}}} = 0.7$, $p^{c|\pi_{z_3}^*}(a) = \frac{\pi_{z_3}^* \Theta_{z_3,a,z_3} \mathscr{I}_{z_3}^c}{\lambda^{c|\pi_{z_3}^*}} = 0.3$, $p^{c|\pi_{z_3}^*}(b) = \frac{\pi_{z_3}^* \Theta_{z_3,b,z_3} \mathscr{I}_{z_3}^c}{\lambda^{c|\pi_{z_3}^*}} = 0.7$.

For z_4 and $\pi^*_{z_4}$, $\lambda^{s|\pi^*_{z_4}} = 1, \lambda^{c|\pi^*_{z_4}} = 0$, $p^{s|\pi^*_{z_4}}(a) = \frac{\pi^*_{z_4}\Theta_{z_4,a,z_4}\mathscr{I}^s_{z_4}}{\lambda^{s|\pi^*_{z_4}}} = 0.3$,

$p^{s|\pi^*_{z_4}}(b) = \frac{\pi^*_{z_4}\Theta_{z_4,b,z_4}\mathscr{I}^s_{z_4}}{\lambda^{s|\pi^*_{z_4}}} = 0.7, p^{c|\pi^*_{z_4}}(a) = p^{c|\pi^*_{z_4}}(b) = 0$.

5. Then, we have

$$\lim_{n\to\infty} D(p^s_n, p^c_n) = H(\{\lambda^s, \lambda^c\})$$
$$+ \sum_{z:z \text{ is recurrent}} p(z)[-H(\{\lambda^{s|\pi^*_z}, \lambda^{c|\pi^*_z}\}) + D(p^{s|\pi^*_z}, p^{c|\pi^*_z})]$$
$$= 0.6043.$$

Thus, for the system in Fig. 2, the worst case secrecy loss, as measured by the limiting JSD, is **0.6043**.

Application to Cache Side-Channel Attack. For the cache side-channel attack model of Fig. 1b, the observer model is given in Fig. 1c. It can be computed that $p(z_1) = 1/6$, $p(z_2) = 5/6$, $\pi^*_{z_1} = [1]$, $\pi^*_{z_2} = [0.6 \quad 0.4]$, $\lambda_s = 1/3$ and $\lambda_c = 2/3$. From which, the limiting divergence $\lim_{n\to\infty} D(p^s_n, p^c_n) = \mathbf{0}$, meaning that no amount of secrecy could be leaked through the side-channel if the cache line is periodically evicted by the processor.

5 Quantification of Distributed Secrecy Loss in Stochastic PODESs Under Bounded-Delay Communications

We now extend the analysis of previous section to study the secrecy quantification in stochastic PODESs in the presence of distributed collusive attackers/observers, each with its own local partial observability, and where the local observers collude and exchange their observations over communication channels with bounded delays, to be able to infer more about the system secrets.

d-**Delaying&Masking Communication Channel**. Figure 3a shows the architecture of a system with distributed observers/attackers, where it is assumed for simplicity and without loss of any generality that there are two local observers at two local sites $I = \{1,2\}$. Each site has three modules [17]: (i) observation mask $M_i : \overline{\Sigma} \to \overline{\Delta}_i$, where Δ_i is the set of locally observed symbols and $M_i(\varepsilon) = \varepsilon$ (M_i can be extended to Σ^* as follows: $M_i(\varepsilon) = \varepsilon$, and $\forall s \in \Sigma^*, \sigma \in \overline{\Sigma}, M_i(s\sigma) = M_i(s)M_i(\sigma)$), (ii) communication channels $C^{(d)}_{ij}, j \neq i, i,j \in I$, which are lossless and order-preserving, but introduce delays bounded by d, and (iii) observer Obs_i, that tracks the system "information-state" following the arrival of its local observations and the communicated observations received from other sites $j \in I, j \neq i$.

The communication channel is a "*delay-block*" with d-bounded communication delay that holds the transmitted information in First-In-First-Out (*FIFO*) manner for at most d delay steps. Accordingly, since there can be at most d events executed by

(a) **(b)** $\mathcal{G}^{(d)} = G\|C_{12}^{(d)}\|C_{21}^{(d)}$

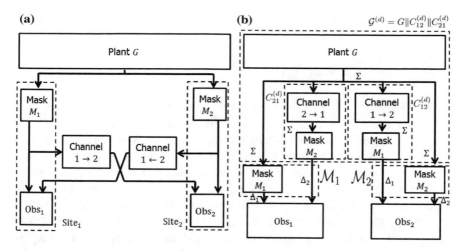

Fig. 3 a Distributed secrecy system architecture to **b** equivalent system architecture (reproduced from [17])

system G between the transmission and the reception of a message on a channel, the channel has a maximum queue length $d+1$. Also, the channel queue evolves whenever a system event occurs, or a transmitted observation is delivered to a destination observer, where such arrival and departure events occur asynchronously. Accordingly, the d-delaying&masking non-stochastic channel model from site-i to site-j $(i \neq j, i, j \in I)$ is of the form, $C_{ij}^{(d)} = (Q_{ij}^{(d)}, \Sigma \cup \overline{\Delta}_i, \beta_{ij}^{(d)}, q_0)$, with the elements as follows. $Q_{ij}^{(d)} \subseteq \Sigma^*$ denotes the set of states, which are the event traces executed in the system but their observed values pending to be delivered at the destination. For $q \in Q_{ij}^{(d)}$, it holds that $|q| \leq d+1$. $\Sigma \cup \overline{\Delta}_i$ is the event set of $C_{ij}^{(d)}$, where Σ is its set of input events and $\overline{\Delta}_i$ is its set of output events. Without loss of generality, we assume that $\Sigma \cap \overline{\Delta}_i = \emptyset$, and $\Delta_i \cap \Delta_j = \emptyset, (j \neq i)$ (otherwise, we can simply rename some of the symbols). $q_0 = \varepsilon$ is the initial state, whereas the transition function $\beta_{ij}^{(d)}$ is defined as follows:

1. "Arrival" due to an event execution in the system: $\forall q \in Q_{ij}^{(d)}, \forall \sigma \in \Sigma$, if $|q| \leq d$, then $\beta_{ij}^{(d)}(q, \sigma) = q\sigma$,

2. "Departure" due to a reception at the destination observer: $\forall q \in Q_{ij}^{(d)}, \forall \delta_i \in \overline{\Delta}_i$, if $M_i(head(q)) = \delta_i$, then $\beta_{ij}^{(d)}(q, \delta_i) = q\backslash head(q)$,

3. Undefined, otherwise,

where $head(q)$ is the first event in trace q, and the after operator "\" in $q\backslash head(q)$ returns the trace after removing the initial event $head(q)$ from the trace q.

Example 3 A system model G is shown in Fig. 4a, with $L(G) = a^+ \cup ba^* \cup uba^+$. Suppose the observation masks of two local sites are defined as follows:

- $M_1(a) = a'$, $M_1(b) = M_1(u) = \varepsilon$, and
- $M_2(b) = b'$, $M_2(a) = M_2(u) = \varepsilon$.

For delay $d = 0$, Fig. 4b shows the model $C_{12}^{(0)}$, and for *delay $d = 1$*, Fig. 4c, d show the models $C_{12}^{(1)}$ and $C_{21}^{(1)}$, respectively. If we follow the trace bab' in $C_{21}^{(1)}$, the states ε, b, ba and a are traversed sequentially. This corresponds to the situation in which site-2 sends out its observation b' to site-1 following the execution of ba in the system, whereas the observation of event a is pending to be received at site-1.

Next, since the operations of masking and delaying can be interchanged, the behaviors under the schematic of Fig. 3a are equivalent to those of Fig. 3b. Then, it is clear that the distributed setting of Fig. 3a can be converted to a decentralized setting of Fig. 3b, having an extended system $\mathscr{G}^{(d)}$ and local observers having the extended observation masks $\{\mathscr{M}_i\}$, defined below. The extended system is given by $\mathscr{G}^{(d)} = G\|_{i,j \in I, i \neq j} C_{ij}^{(d)}$, whereas the extended system model $\mathscr{G}_i$ at site-i $(i \in I)$ includes the system model and only the incoming channel models: $\mathscr{G}_i = G\|_{j \in I - \{i\}} C_{ji}^{(d)}$. The *extended system* $\mathscr{G}_i$ "generates" events in $\Sigma \cup_{j \neq i} \Delta_j$, which are observed by site-i observer Obs_i through an extended observation mask $\mathscr{M}_i : \overline{\Sigma} \cup_{j \in I - \{i\}} \overline{\Delta}_j \to \overline{\Delta} = \cup_{i \in I} \overline{\Delta}_i$. $\mathscr{M}_i$ acts the same as M_i for events in Σ, whereas it is an identity mask for events in Δ_j $(j \neq i)$. Formally, it is defined as follows:

$$\mathscr{M}_i(\sigma) := \begin{cases} M_i(\sigma), & \sigma \in \Sigma, \\ \sigma, & \sigma \in \Delta_j \ (j \neq i). \end{cases} \tag{4}$$

The extended system model at site-i $(i \in I)$ can be refined with respect to the specification to identify the secret and cover behaviors as *states* in the refined system, and is given by $\mathscr{G}_i^R = G\|_{j \in I - \{i\}} C_{ji}^{(d)} \| \overline{R}$.

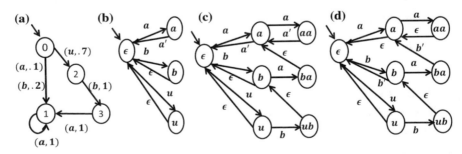

Fig. 4 **a** Stochastic PODES G, **b** $C_{12}^{(0)}$, **c** $C_{12}^{(1)}$, **d** $C_{21}^{(1)}$

Next, we assign probabilities to transitions in $\mathscr{G}_i^R$ as follows. For each state in $\mathscr{G}_i^R$, the transition is either one of the system events, or at most one of channel j $(j \neq i)$ events (either arrival or departure of that channel). Suppose at a system $\mathscr{G}_i^R$ state, with vector of all incoming channel lengths $\mathbf{k}$, the system event is picked with probability $p_{\mathbf{k}}^0$, and suppose the channel j $(j \neq i)$ event can occur with probability $p_{\mathbf{k}}^j$ such that, $p_{\mathbf{k}}^0 + \sum_{j \neq i} p_{\mathbf{k}}^j = 1$. We also require that when all channels are empty $(\mathbf{k} = \mathbf{0})$, $p_{\mathbf{k}}^0 = 1$ (so no channel output can occur when channels are empty), when all channels are full $(\mathbf{k} = \overrightarrow{d+1})$, $p_{\mathbf{k}}^0 = 0$ (so no channel input can occur when channels are full), and if channel j has higher queue length than channel j' $(\mathbf{k_j} \geq \mathbf{k_j'})$, then it can be expected that $p_{\mathbf{k}}^j \geq p_{\mathbf{k}}^{j'}$ (channel j event is more likely than channel j' event when channel j has more number of pending observations). With this choice of selection probability of events, refined extended system model is given by $\mathscr{G}_i^R =$ $(X \times (\Pi_{j \neq i} Q_{ji}^{(d)}) \times \overline{Y}, \Sigma \cup_{j \neq i} \overline{\Delta}_j, \gamma, (x_0, \mathbf{q}_0, y_0))$, where $\overline{Y} = Y \cup \{D\}$, and $\forall (x, \mathbf{q}, \bar{y})$, $(x', \mathbf{q}', \bar{y}') \in X \times (\Pi_{j \neq i} Q_{ji}^{(d)}) \times \overline{Y}$, $\sigma \in \Sigma \cup_{j \neq i} \overline{\Delta}_j$,

$$\gamma((x, \mathbf{q}, \bar{y}), \sigma, (x', \mathbf{q}', \bar{y}')) = \begin{cases} \alpha(x, \sigma, x') \times p_{\mathbf{k}}^0 & \text{if } \sigma \in \Sigma, \\ p_{\mathbf{k}}^j & \text{if } \sigma \in \cup_{j \neq i} \overline{\Delta}_j, \end{cases}$$

if the following holds:

$$(\bar{y}, \bar{y}' \in Y \wedge \beta(\bar{y}, \sigma, \bar{y}') > 0) \vee (\bar{y} = \bar{y}' = D)$$
$$\vee (\bar{y}' = D \wedge \sum_{y \in Y} \beta(\bar{y}, \sigma, y) = 0),$$

and otherwise, $\gamma((x, \mathbf{q}, \bar{y}), \sigma, (x', \mathbf{q}', \bar{y}')) = 0$.

The computation of an observer transition structure for $\mathscr{G}_i^R$ and the associated transition matrices $\{\Theta(\delta)|\delta \in \Delta\}$, is exactly the same as in the centralized setting, as is described in the appendix.

Example 4 Continuing Example 3, suppose the delay bound $d = 1$, so there are three possibilities for the length of the only channel, $\mathbf{k} = \{0, 1, 2\}$. Let $p_0^0 = 1, p_1^0 = 0.5, p_2^0 = 0$ (implying $p_0^2 = 1 - p_0^0 = 0$, $p_1^2 = 1 - p_1^0 = 0.5$, $p_2^2 = 1 - p_2^0 = 1$). Figure 5a shows the extended system model $\mathscr{G}_1$ at site-1. Suppose R is given in Fig. 5b, i.e., $K = L(R) = a^+ \cup ba^*$. Then, the refinement G_1^R is shown in Fig. 5c. So for example, at the initial state $(0, \varepsilon, 0)$, the channel is empty, and no channel events occur at this state $(p_0^2 = 0$ while $p_0^0 = 1)$. Then, for any system event $\sigma \in \Sigma$, $\gamma((0, \varepsilon, 0), u, (2, u, D)) = \alpha(0, u, 2) \times p_0^0 = 0.7 \times 1 = 0.7, \gamma((0, \varepsilon, 0), b, (1, b, 1)) = \alpha(0, b, 1) \times p_0^0 = 0.2 \times 1 = 0.2$, and $\gamma((0, \varepsilon, 0), a, (1, a, 1)) = \alpha(0, a, 1) \times p_0^0 = 0.1 \times 1 = 0.1$. Whereas, at state $(2, u, D)$, there is observation u queued up in the channel. Thus, either the system can execute a new event $b \in \Sigma$, with probability $\gamma((2, u, D), b, (3, ub, D)) = \alpha(2, b, 3) \times p_1^0 = 1 \times p_1^0 = 0.5$, or a channel event can occur, with probability $\gamma((2, u, D), \varepsilon, (2, \varepsilon, D)) = p_1^2 = 0.5$. The remaining state

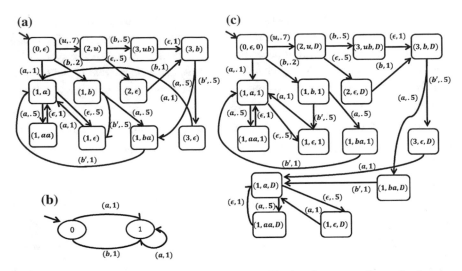

Fig. 5 a Extended system model $\mathcal{G}_1$ at site-1, **b** specification for secrets, R, **c** refined system model $\mathcal{G}_1^R$

transitions can be computed similarly. The models $\mathcal{G}_2$ and $\mathcal{G}_2^R$ at site-2 can be generated in a manner similar to $\mathcal{G}_1$ and $\mathcal{G}_1^R$, respectively.

In Sect. 4, we presented a way to compute JSD-based measure of secrecy loss for stochastic PODES when there is a single observer. To compute the secrecy loss in the distributed setting, resulting from the aggregated observations at any site-i ($i \in I$), which include it's own immediate observations and the delayed communicated observations from other distributed sites, the JSD computation can be carried out over the refined extended system model $\mathcal{G}_i^R$, following the method introduced in Sect. 4. The example below illustrates the extended observer structure and the corresponding JSD based secrecy loss computation in a distributed collusive setting, respectively.

Example 5 Consider the refined extended system model of Fig. 5c at site-1 where $\mathcal{M}_1(a) = a'$, $\mathcal{M}_1(b) = \mathcal{M}_1(u) = \varepsilon$, while the extended mask function is the identity function over the received observations, $\Delta_2 = \{b'\}$. Then, Fig. 6a shows the extended observer Obs_1.

Then, based on Obs_1, the following computation illustrates the steps of JSD computation at site-1.

1. $\sum_z |z| = 14$ and so $\tilde{\Theta}$ is a 14×14 matrix with entries:
   $\tilde{\Theta}(1,2) = \tilde{\Theta}(1,3) = \tilde{\Theta}(1,5) = 0.1, \quad \tilde{\Theta}(1,4) = \tilde{\Theta}(1,6) = 0.35, \quad \tilde{\Theta}(2,7) = \tilde{\Theta}(2,8) = \tilde{\Theta}(7,7) = \tilde{\Theta}(7,8) = \tilde{\Theta}(8,7) = \tilde{\Theta}(8,8) = \tilde{\Theta}(9,11) = \tilde{\Theta}(9,12) = \tilde{\Theta}(10,13) = \tilde{\Theta}(10,14) = \tilde{\Theta}(11,11) = \tilde{\Theta}(11,12) = \tilde{\Theta}(12,11) = \tilde{\Theta}(12,12) =$

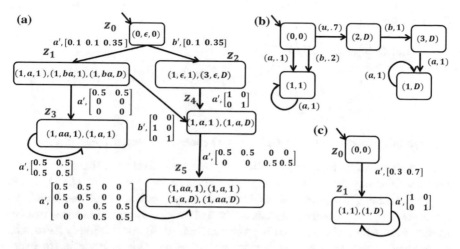

Fig. 6 **a** Observer Obs_1 for the system of Fig. 5c, **b** model G^R for system of Fig. 4a under no collusion, **c** observer under no collusion

$\tilde{\Theta}(13, 13) = \quad \tilde{\Theta}(13, 14) = \tilde{\Theta}(14, 13) = \tilde{\Theta}(14, 14) = 0.5, \quad \tilde{\Theta}(3, 9) =$
$\tilde{\Theta}(4, 10) = \tilde{\Theta}(5, 9) = \tilde{\Theta}(6, 10) = 1$ and zeros elsewhere.

Then, $\qquad \qquad \qquad \qquad \pi^* = [0 \quad 0 \quad 0 \quad 0 \quad 0 \quad 0 \quad 0$
$0.05 \quad 0.05 \quad 0 \quad 0 \quad 0.1 \quad 0.1 \quad 0.35 \quad 0.35]. \qquad \qquad$ Therefore,
$p(z_0) = p(z_1) = p(z_2) = 0, p(z_3) = 0.1, p(z_4) = 0,$ and $p(z_5) = 0.9.$

2. Here $\qquad \mathscr{I}^s = [1 \quad 1 \quad 1 \quad 0 \quad 1 \quad 0 \quad 1 \quad 1 \quad 1 \quad 0 \quad 1 \quad 1 \quad 0 \quad 0]^T,$
$\mathscr{I}^c = [0 \quad 0 \quad 0 \quad 1 \quad 0 \quad 1 \quad 0 \quad 0 \quad 0 \quad 1 \quad 0 \quad 0 \quad 1 \quad 1]^T.$ And so $\lambda^s = 0.3$
and $\lambda^c = 0.7.$

3. Here z_3, and z_5 are recurrent nodes, and each of them forms a SCC. We have $\pi^*_{z_3} = [0.5 \quad 0.5]$, and while there are multiple solutions to the equation set $\pi^*_{z_5} = \frac{\pi^*_{z_5} \Theta_{z_5, a' z_5}}{\|\pi^*_{z_5} \Theta_{z_5, a' z_5}\|}$, only $\pi^*_{z_5} = [0.11 \quad 0.11 \quad 0.39 \quad 0.39]$ is reachable. Thus, each set of recurrent nodes is a singleton set, and each with a unique fixed-point distribution. Therefore, for each recurrent node z, $p(z, k) = p(z)$.

4. Here $\qquad \mathscr{I}^s_{z_3} = [1 \quad 1]^T, \qquad \mathscr{I}^c_{z_3} = [0 \quad 0]^T, \qquad \mathscr{I}^s_{z_5} = [1 \quad 1 \quad 0 \quad 0]^T,$
$\mathscr{I}^c_{z_5} = [0 \quad 0 \quad 1 \quad 1]^T.$ For z_3 and $\pi^*_{z_3}$, $\lambda^{s|\pi^*_{z_3}} = 1, \lambda^{c|\pi^*_{z_3}} = 0, p^{s|\pi^*_{z_3}}(a') =$
$\frac{\pi^*_{z_3} \Theta_{z_3, a' z_3} \mathscr{I}^s_{z_3}}{\lambda^{s|\pi^*_{z_3}}} = 1, p^{s|\pi^*_{z_3}}(b') = p^{c|\pi^*_{z_3}}(b') = p^{c|\pi^*_{z_3}}(a') = 0.$ For z_5 and $\pi^*_{z_5}$, $\lambda^{s|\pi^*_{z_5}} =$
$0.22, \lambda^{c|\pi^*_{z_5}} = 0.78, \quad p^{s|\pi^*_{z_5}}(a') = \frac{\pi^*_{z_5} \Theta_{z_5, a' z_5} \mathscr{I}^s_{z_5}}{\lambda^{s|\pi^*_{z_5}}} = 1, \quad p^{c|\pi^*_{z_5}}(a') = \frac{\pi^*_{z_5} \Theta_{z_5, a' z_5} \mathscr{I}^c_{z_5}}{\lambda^{c|\pi^*_{z_5}}} = 1,$
$p^{s|\pi^*_{z_5}}(b') = p^{c|\pi^*_{z_5}}(b') = 0.$

5. Then, we have

$$\lim_{n \to \infty} D_1(p_n^s, p_n^c) = H(\{\lambda^s, \lambda^c\})$$
$$+ \sum_{z:z \text{ is recurrent}} p(z)[-H(\{\lambda^{s|\pi_z^*}, \lambda^{c|\pi_z^*}\}) + D_1(p^{s|\pi_z^*}, p^{c|\pi_z^*})]$$
$$= \mathbf{0}.197.$$

Note this happens to be the same as JSD measure of secrecy loss at site-2.

In contrast, when there is no collusion among observers (so there is no communication among the two sites), Fig. 6b, c show, respectively, the refined system G^R (no incoming channels and so identical refined model at all sites) and the corresponding site-1 observer structure. The JSD value, computed in same manner as above but with respect to the observer structure of Fig. 6c, is simply **Zero**, i.e., no amount of secrets is revealed under no collusion. This is because for every observation, the probability of it coming from secrets in K vs from covers in $L - K$ is exactly the same.

6 Conclusion

In this chapter, we presented information theoretic measure for secrecy loss quantification in PODESs in both centralized versus distributed collusive settings, in the presence of a single attacker/observer versus multiple attackers/observers exchanging their observations, respectively. The statistical difference, in the form of the Jensen-Shannon Divergence, between the influence of secrets versus covers on the observations, is employed to quantify the loss of secrecy. It is shown that this JSD measure is equivalent to the mutual information between the distribution over the possible observations versus that over the possible status of system execution (whether secret or cover). An observer structure is formed and used to aide the computation of JSD in the limit as the length of the observation approaches infinity to quantify the worst case loss of secrecy. In distributed collusive setting, channel models are introduced to extend the system model to capture the effect of exchange of observations, and the JSD computation of the centralized case is applied over the extended model to arrive at the measure for secrecy loss. Future work will involve developing a software tool for JSD computation, and performing application studies. Knowing the JSD value can help an engineer to perform secrecy analysis of a system, and revisit the system design to make it improve its level of secrecy as needed.

Acknowledgments This research was supported in part by Security and Software Engineering Research Center (S2ERC), and the National Science Foundation under the grants NSF-CCF-1331390 and NSF-ECCS 1509420.

Appendix

In this appendix, we describe the computation of an observer transition structure that can be used to track the evolution of G^R over its observed symbols Δ, and the associated transition matrices $\{\Theta(\delta) | \delta \in \Delta\}$. Given the refined system model G^R, and its observation mask $M : \overline{\Sigma} \to \overline{\Delta}$, define the set of traces originating at $(x, \bar{y})$, terminating at $(x', \bar{y}')$ and executing a sequence of unobservable events followed by a single observable event with observation δ as $L_{G^R}((x, \bar{y}), \delta, (x', \bar{y}')) := \{s \in \Sigma^* | s = u\sigma, \ M(u) = \varepsilon, M(\sigma) = \delta, \ \gamma((x, \bar{y}), s, (x', \bar{y}')) > 0\}$. Define its probability, $\alpha(L_{G^R}((x, \bar{y}), \delta, (x', \bar{y}'))) := \sum_{s \in L_{G^R}((x,\bar{y}),\delta,(x',\bar{y}'))} \gamma((x, \bar{y}), s, (x', \bar{y}'))$, and denote it as $\theta_{(x,\bar{y}),\delta,(x',\bar{y}')}$. Also, define $\lambda_{ij} = \sum_{\sigma \in \Sigma_{uo}} \gamma(i, \sigma, j)$ as the probability of transitioning from $(x, \bar{y})$ to $(x', \bar{y}')$ while executing a single unobservable event. Then, letting $i = (x, \bar{y})$ and $j = (x', \bar{y}')$, $\theta_{i,\delta,j} = \sum_k \lambda_{ik} \theta_{k,\delta,j} + \sum_{\sigma \in \Sigma:M(\sigma)=\delta} (i, \sigma, j)$, where the first term on the right hand side (RHS) corresponds to transitioning in at least two steps (i to intermediate k unobservably, and k to j with a single observation δ at the end), whereas the second term on RHS corresponds to transitioning in exactly one step [3, 12]. Thus, for each $\delta \in \Delta$, all the probabilities $\{\theta_{i,\delta,j} | i, j \in X \times \overline{Y}\}$ can be found by solving the following matrix equation [23]: $\Theta(\delta) = \Lambda\Theta(\delta) + \Gamma(\delta)$, where $\Theta(\delta), \Lambda$ and $\Gamma(\delta)$ are all $|X \times \overline{Y}| \times |X \times \overline{Y}|$ square matrices whose *ijth* elements are given by $\theta_{i,\delta,j}, \lambda_{ij}$ and $\sum_{\sigma \in \Sigma:M(\sigma)=\delta} \gamma((x, \bar{y}), \sigma, (x', \bar{y}'))$, respectively.

References

1. Backes, M., Köpf, B., Rybalchenko, A.: Automatic discovery and quantification of information leaks. In: Proceedings of 30th IEEE Symposium on Security and Privacy, pp. 141–153, Washington, DC, 2009
2. Bryans, J., Koutny, M., Mu, C.: Towards quantitative analysis of opacity. Technical Reports Series, Newcastle University (2011)
3. Chen, J., Ibrahim, M., Kumar, R.: Quantification of secrecy in partially observed stochastic discrete event systems. IEEE Trans. Autom. Sci. Eng. (accepted (Sept. 2015))
4. Chen, J., Kumar, R.: Failure detection framework for stochastic discrete event systems with guaranteed error bounds. IEEE Trans. Autom. Control **60**(6), 1542–1553 (2015)
5. Christian, S., Collberg, C.T.: Watermarking, tamper-proofing, and obfuscation-tools for software protection. IEEE Trans. Softw. Eng. **28**(8), 735–746 (2002)
6. Cover, T.M., Thomas, J.A.: Elements of information theory. Wiley, New York (2012)
7. Daemen, J., Rijmen, V.: Aes proposal: Rijndael, version 2, aes submission (1999)
8. Espinoza, B., Smith, G.: Min-entropy as a resource. Inf. Comput. **226**, 57–75 (2013)
9. Garg, S., Gentry, C., Halevi, S., Raykova, M., Sahai, A., Waters, B.: Candidate indistinguishability obfuscation and functional encryption for all circuits. In: IEEE 54th Annual Symposium on Foundations of Computer Science (FOCS' 2013), pp. 40–49, Berkeley, CA, 2013
10. Garg, V.K., Kumar, R., Marcus, S.I.: A probabilistic language formalism for stochastic discrete-event systems **44**(2), 280–293 (1999)

11. Ibrahim, M., Chen, J., Kumar, R.: Secrecy in stochastic discrete event systems. In: Proceedings of 11th IEEE International Conference on Networking, Sensing and Control (ICNSC'14), pp. 48–53. Miami, FL, 2014
12. Ibrahim, M., Chen, J., Kumar, R.: An information theoretic measure for secrecy loss in stochastic discrete event systems. In: Proceedings of the 7th International Conference on Electronics, Computers and Artificial Intelligence (ECAI'15), pp. 1–6, Bucharest, 2015
13. Jacob, R., Lesage, J.J., Faure, J.M.: Opacity of discrete event systems: models, validation and quantification. In: Proceedings of the 5th International Workshop on Dependable Control of Discrete Systems (DCDS'15), hal-01139890, Cancun, Mexico, 2015
14. Kaijser, T.: A limit theorem for partially observed markov chains. Ann. Prob. **3**(4), 677–696 (1975)
15. Kak, A.: Aes: Lecture notes in computer and network security (2015). Purdue University, https://engineering.purdue.edu/kak/compsec/NewLectures/Lecture8.pdf. Accessed 1 May 2015
16. Kundur, D., Ahsan, K.: Practical internet steganography: Data hiding in ip. In: Proceedings of Texas Workshop on Security of Information Systems, College Station, Texas, 2003
17. Qiu, W., Kumar, R.: Distributed diagnosis under bounded-delay communication of immediately forwarded local observations. IEEE Trans. Syst. Man Cybern. Part A: Syst. Humans (2008)
18. Ren, J., Wu, J.: Survey on anonymous communications in computer networks. Comput. Commun. **33**, 420–431 (2010)
19. Saboori, A., Hadjicostis, C.N.: Opacity verification in stochastic discrete event systems. In: Proceedings of 49th IEEE Conference on Decision and Control, pp. 6759–6764, Atlanta, GA, 2010
20. Saboori, A., Hadjicostis, C.N.: Probabilistic current-state opacity is undecidable. In: Proceedings of 19th International Symposium on Mathematical Theory Network and Systems (MTNS '2010), pp. 477–483, Budapest, Hungary, 2010
21. Smith, G.: On the foundations of quantitative information flow. In: Proceedings of International Conference on Foundations of Software Science and Computation Structures (FoSSaCS 09), pp. 288–302, 2009
22. Takai, S., Kumar, R.: Verification and synthesis for secrecy in discrete-event systems. In: Proceedings of IEEE American Control Conference, (ACC '09), pp. 4741–4746, St. Louis, MO, 2009
23. Wang, X., Ray, A.: A language measure for performance evaluation of discrete-event supervisory control systems. Appl. Math. Model. **28**(9), 817–833 (2004)
24. Xie, A., Beerel, P.A.: Efficient state classification of finite-state Markov chains. IEEE Trans. Comput. Aided Des. Integr. Circuits Syst. **17**(12), 1334–1339 (1998)
25. Zhang, T., Lee, R.B.: Secure cache modeling for measuring side-channel leakage. Technical Report, Princeton University (2014)

Framework for Cognitive Radio Deployment in Large Scale WSN-UAV Surveillance

Grigore Stamatescu, Dan Popescu and Radu Dobrescu

Abstract The chapter discusses the integration of heterogenous entities operating together to achieve common mission objectives that have been increasingly adopted for monitoring and surveillance of interest areas and physical infrastructures, under an intelligent environment paradigm. Several technologies have been adopted recently which include, wireless sensor networks (WSN), terrestrial remote operated vehicles (ROV) and unmanned aerial vehicles (UAV). In this context, a multi-level system framework for multi-sensory robotic surveillance of critical infrastructure protection through communication, data acquisition and processing is introduced and discussed. Cognitive radio (CR) is seen as a key enabler of the multi-level framework communication, arguing that by exploiting in an opportunistic fashion the time, frequency and spatial stream of the wireless environment, increased communication reliability can be achieved with positive impact on the availability and service level at each hierarchical level. By classification and pointing out the key advantages of CR in such a scenario, we underline the advantages of this scheme within the constraints of a working scenario based on multi-level ground sensor network and aerial robotic platform integration.

Keywords Wireless sensor networks · Unmanned aerial vehicles · Information processing · Data acquisition · Cognitive radio · Large scale surveillance

G. Stamatescu (✉) · D. Popescu · R. Dobrescu
Department of Automatic Control and Industrial Informatics,
University "Politehnica" of Bucharest, 313 Spl. Independentei,
060042 Bucharest, Romania
e-mail: grigore.stamatescu@upb.ro

D. Popescu
e-mail: dan.popescu@upb.ro

R. Dobrescu
e-mail: radu.dobrescu@upb.ro

© Springer International Publishing Switzerland 2016
E. Pricop and G. Stamatescu (eds.), *Recent Advances in Systems Safety and Security*,
Studies in Systems, Decision and Control 62, DOI 10.1007/978-3-319-32525-5_3

1 Introduction

Dense instrumentation of the physical world, mainly through networks of cooperating objects [1] has lead to the emergence of the new intelligent environments paradigm [2]. In these types of scenarios, various autonomous fixed or mobile entities collaborate in order to assure achieving specific objectives, enhancing the human factor for dependable, safe and secure systems. One example of a relevant application has been BorderSense [3], leveraging multiple types of wireless sensor networks, including: multimedia and underground to build a full system for real-time border surveillance. At the top layer of the proposed system, lay unmanned autonomous aerial vehicles which act as high level sensory platforms for complex imaging and high data link communication capacity. This showcases the growing body of research at the interface between WSN and UAV [4]. It is argued that by exploiting the advantages of each platform, the individual drawbacks can be mitigated in achieving superior performance from the system as a whole. Not only critical applications can be enhanced through such approaches but rather the field can be extended towards other types of monitoring and surveillance scenarios in civil [5], commercial and industrial applications.

Approaching such complex issues from a communication stand point, there is an increasing interest in the use of relatively small, flexible unmanned aerial vehicles (UAVs) that fly at lower altitude for providing relay services for mobile ad hoc networks with ground-based communication nodes. The UAV acts as a decode-and-forward relay, sending the messages from the co-channel users on the ground to some remote base station.

A number of different approaches have been proposed in the literature to address the performance of UAV-assisted communication networks. For example, in [6], a throughput maximization protocol for non-real time applications was proposed for a network with UAV relays in which the UAV first loads data from the source node and then flies to the destination node to deliver it. The authors in [7] investigated different metrics for ad hoc network connectivity and propose several approaches for improving the connectivity through deployment of a UAV. In [8], the authors considered a scenario in which multiple UAVs are deployed to relay data from isolated ground sensors to a base station, and an algorithm was proposed to maintain the connectivity of the links between the sensors and base station.

The work described above assumes that the ground nodes are static and that the UAV is configured with only a single communication channel, but given the benefits of employing multiple channels for communications, other authors have considered the advantages they offer for UAV-based platforms. A swarm of single antenna UAVs was used as a virtual antenna array to relay data from a fixed ad hoc network on the ground in [9]. A relay system with multi-antenna UAVs and multi-antenna mobile ground terminals was investigated in [10].

The system of air-ground communication is one of the most fundamental elements of the surveillance system proposed in this paper. In the last years, due to the increasing of the number of data applications for such communication, the demand to effectively use the limited frequency spectrum has increased. Given the limitations of the natural frequency spectrum, it is obvious that the current static frequency allocation schemes cannot meet the requirements of the air-to-ground bidirectional communication [11]. There are several proposed approaches to solve the drawbacks of static spectrum allocation, which is the major bottleneck for effective use of the limited spectrum, but the inefficiency in the spectrum usage is the new communication paradigm of Dynamic Spectrum Access (DSA) [12], able to exploit wireless spectrum opportunities.

The key enabling technology of DSA is the Cognitive Radio (CR) technology that is built on a software-defined radio (SDR). SDR technology allows multimode, multi-band and/or multi-functional wireless devices that can be enhanced using software upgrades. Simply SDR is defined as "Radio in which some or all the physical layer functions are software defined" [13]. As component of SDR, CR is a part of the Adaptive Radio, radio in which communications systems have a means of monitoring their own performance and modifying their operating parameters to improve performance, and at its turn has as inner core the Intelligent Radio technology, which allows the cognitive radio to improve the ways in which it adapts to changes in performance and environment [14].

Therefore, CR is defined as an intelligent wireless communication system that is aware of its environment and uses this methodology to learn from the environment and adapt to statistical variations in the input stimuli. It provides the opportunity to address the static allocation of spectrum issue and offer a more flexible transition approach for the control of an air-ground radio system.

In this chapter, a cognitive radio (CR) scheme is discussed as key point in exploiting in optimal way the time, frequency and spatial stream of the wireless environment, with positive impact on the communication performance in a multi-level monitoring and surveillance framework. The rest of the paper is structured as follows. Section 2 frames our contribution into the context of related worked, previously handling the emergence of cognitive radio wireless sensor networks (CR-WSNs). Section 3 describes a new framework and system architecture for heterogeneous multi-level monitoring based on WSN and UAV, focusing on the communication as background for cognitive radio implementation and leveraging its advantages. Cooperative spectrum sensing methods for CR within this scenario are discussed in Sect. 4. As a means to evaluate and test the proposed approach, methods for implementation of a CR testbed with this goal, are given in Sect. 5. Final remarks and outlook on future work are handled in Sect. 6.

The chapter represents an improved and extended version of the earlier conference contribution in [15]. More detailed related work background and operational details regarding the cooperative WSN-UAV sensing and communication frameworks are given.

2 Related Work

Application of cognitive radio to embedded sensor networks was initially discussed in [16] in the form of an extended review. The main spectrum sensing methods are identified as follows: matched filtering, energy detection, feature detection and interference temperature. Advantages and disadvantages of these methods are described, to be balanced depending on the computational resources available at the individual node level, amount of previous information and constraints on spectrum sensing duration. Subsequently the analysis evaluates the impact that a cognitive radio scheme might have on the communication layers of the wireless sensor network, mainly MAC and routing. At the transport level, the main objective is reliable end-to-end delivery of data while adhering to hard or soft QoS restrictions. Main conclusion that a CR scheme for WSN is viable given the positive perspectives that multi-channel availability brings in specific WSN application scenarios, especially when dealing with bursty communication and unreliable, time variant communication links. At the interface between both CR and WSN domains lay interesting challenges which solved can bring significant benefits for specific applications.

The authors of [17] carry out an analysis of current ISM frequency bands usage in the context of ever increasing need for communication availability. The advantages of CR-WSN are introduced based on a comparison between current ad hoc cognitive radios and current WSN technology. At the hardware level, the generic architecture of a CR-WSN is presented which allows for implementation of logical spectrum management schemes at various complexity levels. It contains an additional dedicated CR unit which acts as an intermediary between the processor and transceiver unit, which is both energy and location aware. Differentiating between signal processing techniques and cooperative sensing for spectrum sensing, the latter category is broken down into: centralized, decentralized and hybrid approaches.

An extended review, albeit for the industrial monitoring and control, use scenarios of software-defined and cognitive radio system is performed in [18]. In [19] wideband spectrum sensing techniques are described while the spectrum sensing unit (SSU) as the essential physical block of a cognitive radio is detailed. The energy efficiency of cooperative spectrum sensing techniques is also analyzed, while concluding that higher cooperative gain of a distributed method has to consider the overall high cost generated by coordination and cooperative overhead upon the network.

3   System Architecture

3.1   *MUROS Framework*

We first introduce the MUROS framework [20] with the purpose to develop a surveillance and monitoring system with unmanned aerial platforms for monitoring, preventing and mitigating incidents which have an impact on critical infrastructure such as transport routes for oil products, railways, power lines, highways etc. The main components of system with unmanned aerial vehicles are illustrated in Fig. 1:

- GCS (ground control station): data and control management;
- GDT (ground data transmission): communication node between GCS and UAV;
- Launcher: launching mechanism for the UAV;
- Sensors: components of the deployable multisensory network;
- UAV: aerial platform which gathers data from sensors and payload.

The originality and innovative character of the UAV implementation within the framework are given by the following elements:

- Automatic launch and landing of the aerial vehicle;
- Solutions for collecting and transmitting high resolution video images to long distances;
- Correlate information collected by the aerial vehicle with the ground sensors network;
- Command center with a flexible architecture which shall allow the possibility to coordinate land and aerial sensors systems;

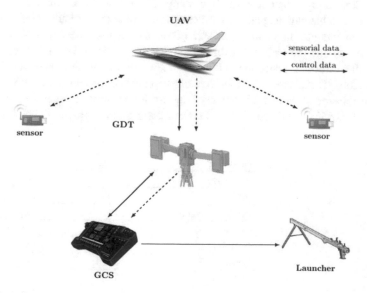

Fig. 1 System components at the UAV level

- Ensure the security of the aerial vehicle by implementing specific recovery algorithms in case of damage.

The main idea is to design a flexible multisensory robotic system capable of monitoring critical infrastructure in a semi-autonomous manner. As it can be seen, the system is composed from:

- a fixed part:
  - ground station;
  - multisensory network.

- a mobile part (the unmanned aerial vehicle—UAV).

The UAV monitors a surveillance area, collects and process data through on-board sensors and from the ground-based sensorial network communicates to and receives instructions from the ground control station. Depending on the particular module, novel developments include control algorithms for trajectory/path tracking, obstacle avoidance/anti-grounding and anti-stall constraints, terrain mapping and coverage, fault tolerant control. Also, acquisition algorithms from UAV and sensorial network, communication algorithms, data processing algorithms such as feature extraction and image analysis are developed.

The current specifications of the UAV subsystem are listed in Table 1. An important mention regarding the operation range is that they depend on the operational mode employed. Operator-based control is available up to 15 km (limited by the data link range) while autonomous mode operation can reach up to 30 km. Figure 2 showcases the conceptual system architecture on a broader scale for WSN-UAV collaboration in common monitoring and surveillance.

The first design option was to employ a multi-level architecture which is modular, scalable and adaptable to the constraints of each particular application. At level 0 we include the central gateway (GW) as the ultimate data collection and control entity and the main point for human operator actions. Level 1 is represented by mobile airborne platforms with advanced communication and sensing e.g. photo/video. These can be either operator-controlled or function autonomously given a mission plan and objectives, support for multiple cooperative operation with other UAVs is included. Level 2 covers the ground deployed WSN with scalar

Table 1 MUROS UAV specifications

Parameter	Value	Unit
Wingspan	4	m
MTOW	10	kg
Payload weight	1	kg
Cruise speed	130	km/h
Operating range	15/30*	km
Service ceiling	3000	m
Autonomy	180	min
Length	1.2	m

*It refers to the extended operation range of the UAV in autonomous mode: "while autonomous mode operation can reach up to 30 km."

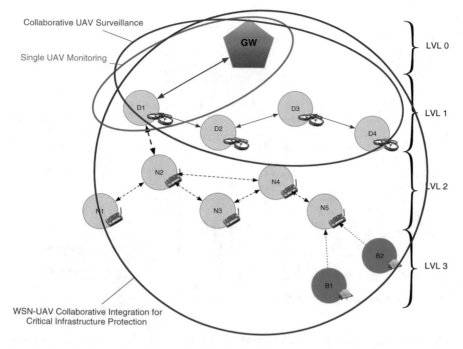

Fig. 2 Heterogenous multi-level architecture for communication, data acquisition and processing —MUROS

sensors e.g. temperature, humidity, magnetic, luminosity, etc., operating in a low power radio mesh network in an energy efficient manner. At the lowest level, level 3, low-cost binary detectors have been included for specific use cases like intrusion detection and target tracking. Based on this four level architecture, three operation modes have been defined, in increasing order of complexity. First, we define the conventional single UAV monitoring of a target area through levels 0 and 1. This extends to multiple UAV collaborative surveillance, when a larger area has to be covered effectively or with different types of airborne sensors and communication links. Finally the most advanced use case encompasses all level from 0 to 3 in a unified platform for WSN-UAV integration.

Functional diagram of the proposed system is shown in Fig. 3. The main elements that are listed in the diagram consist of: L—launcher, GCS—ground control station, GDT—ground data terminal, WSN—wireless sensor network, UAV—unmanned aerial vehicle, NW—WSN component of the UAV system and NCW—central node of the WSN. Functionality is split among the launch phase are preliminary requirement, followed by the operation phase where the communication and cooperation primitives between the WSN and the UAV are deployed. The second distinction is made between the wired communication of the ground-based support infrastructure (L, GDT, GCS, NCW) and wireless communication among the mobile and distributed sensing entities represented by the UAV and WSN.

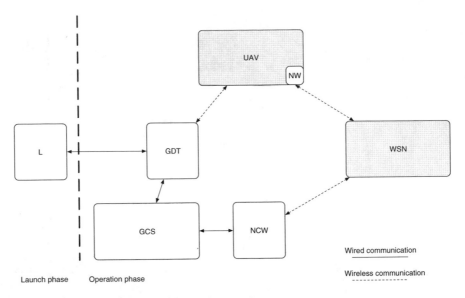

Fig. 3 Functional model of the MUROS framework

Integration at the UAV level is supported by a compatible CR-enabled wireless sensing module which enables interoperability to/from the ground sensor network. Dynamic task allocation within the mission context among the two sensing entities is handled collaboratively.

The GCS is responsible for centralized data aggregation and fusion across the two sensing domains. Information received from the WSN is first handled at the central sensor network node which also has the ability to parameterize the network according to application requirements and to prioritize incoming data. Subsequently, this is forwarded to the GCs where the human operator enters the control loop having the ability to act on the aggregated data. In similar fashion, the UAV data is passed through the GDT and then merged and augmented with the scalar sensor network information at the GCS.

Elaborating on the ground sensor network, with mesh capabilities and low power operation, we define three functional modes:

- normal/low rate data acquisition with low duty cycle 1–2 % of the sensor nodes; data reaches the sensor network sink or they are periodically collected by the UAV and/or mobile terrestrial robotic platforms in the case of partitioned networks; the UAV is operated on demand, given high mission cost and communication link in non-critical scenarios;
- alert level, where certain predefined events are detected through the significant variation of a monitored parameter or by correlation among the reported values of the various on-board sensors; the data acquisition rate grows for a limited timeframe in the interest area, until the system is reset from a superior decision level;

- alarm level, the event is confirmed and continuous monitoring is initiated; this case maximizes the probability of communication channel congestion given the high number of nodes transmitting simultaneously and represents a solid argument in favor of a CR scheme at the communication layer.

3.2 Communication Structure Description as Basis for CR Scheme

Ground stations assure both data collection and event detection. In the basic set-up, their placement is fixed and are linked through a mesh network for both configuration and collaborative information processing towards critical event detection/alerting. The events are reported under real-time constraints. The UAV enhances system functionality by implementing relaying and a communication backchannel. The general organization of a cognitive radio module is illustrated in Fig. 4.

The main separation is between the software defined and the hardware/RF sides. The main role of the software component is to assure best possible utilization and opportunistic switching of the radio channel by leveraging local or networked intelligence. The complexity of the routines is in accordance to the availability of necessary computing resources.

Around each ground station, considered as Primary User (PU) in the CR scheme, an ad hoc CR network is formed with mobile platforms, implemented as ROVs, having the role of transmitting data from the PU towards the UAV. The UAV follows a known route and thus the mobile platforms place themselves as to maximize coverage in the interest area. Even if there is a way to establish a direct PU-UAV link, its role is only to send and receive calibration messages, confirming the degree of confidence for the data transmitted by the intermediary CR stations.

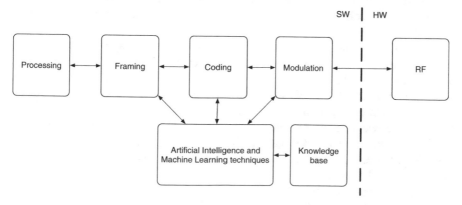

Fig. 4 Cognitive radio organization

Fig. 5 Use case for CR
operation for WSN-UAV
monitoring

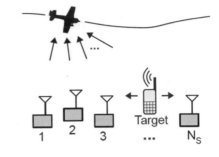

Fig. 6 Communication
subsystem schematic

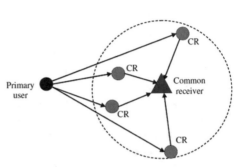

Figures 5 and 6 highlight the fundamental approach for CR communication in
the multi-level monitoring framework. First a use case is suggested based on a
network of CR-equipped ground stations, relaying messages to an aerial UAV
platform, following a predefined route for data collection. CR is essential for
allowing the implementation of real-time constrains on event detection and/or target
tracking, as it leads to efficient usage of the communication channel. The distributed
nature of the communication in the CR scheme is shown next.

We applied the CR based collaborative spectrum sensing in a UAV scenario,
where the UAV serves as a natural collection point for the distributed measurements. The ground nodes in our scheme sense the spectrum in parallel. Each sensor
node transmits measurements when the UAV is in the air in a certain zone around
the ground station.

An additional function of the UAV is to assure the decoding of the received data
sets. By means of the proposed weighted CR scheme, frequencies associated with
higher power signals receive processing priority. Moreover, information detected
from such a priority signal is used to enhance the precision for the decoding of the
information stemming from other sensors.

The proposed scenario assumes the improvement the performance of the communication system. The UAV has sufficient on-board computing resources to carry
out additional processing, such as filtering out invalid results or assigning weights
on the decoded data.

Following the approach from [21], the two main directions for application of the
collaborative decoding have been identified, namely sensor-diversity and

measurement-delivery variation. The former is useful whenever the is a significant difference between signal strengths coming from two or more ground stations at a common receiver. In this way the decoding carried out for the stronger signal(s) can be effectively be applied to improve the outcome of the decoding procedure for the weaker classified stations. The latter case account for the situation in which, at the UAV level, batches of different measurements are received. As these sets might have different sizes, dependant on the individual link characteristics at any given time and emitter-receiver pair, we can use the results from decoding the more significant larger sets upon the smaller ones, with the goal of increased accuracy.

4 Cooperative Spectrum Sensing Scheme for CR

The ultimate objective of the cognitive radio is to obtain the best available spectrum through cognitive capability and reconfigurability. Tasks required for adaptive operation are: Spectrum sensing, spectrum analysis, spectrum decision [22]. Once the operating spectrum band is determined, the communication can be performed over this spectrum band. However, since the radio environment changes over time, space, and frequency, the cognitive radio device should keep track of the changes of the radio environment. If the current spectrum band in use becomes unavailable, the spectrum mobility function is performed to provide a seamless transmission. Any environmental change during the transmission, such as primary user appearance, user movement, or traffic variation, can trigger this adjustment. The main tasks of CR in cognitive cycle are summarized in Fig. 7.

In this paper we mainly focus on the spectrum sensing since it is the most crucial part of the cognitive cycle. One of the great challenges of implementing spectrum sensing is the hidden terminal problem, which occurs when the cognitive radio is shadowed, while a primary user (PU) is operating in the vicinity. In order to deal

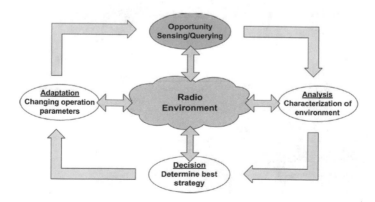

Fig. 7 Simplified cognitive cycle (after [28])

with this problem, multiple cognitive users can cooperate to conduct spectrum sensing.

Let consider a cognitive radio network (CRN) composed of a primary user (PU), N cognitive radios (secondary users SU) CRi (i = 1, ..., N) and a common receiver. The common receiver functions as a base station (BS) which manages the cognitive radio network and all associated N cognitive radios. We assume that each CR performs local spectrum sensing independently, by deciding between the following two hypotheses [23]:

$$
\begin{aligned}
&\text{H0: yi(t)} = \text{ni(t)},\quad \text{if PU is absent}\\
&\text{H1: yi(t)} = \text{his(t)} + \text{ni(t)},\quad \text{if PU is present}
\end{aligned}
\tag{1}
$$

where yi(t) is the observed signal at the ith CR, s(t) is the PU signal assumed to be with zero mean and variance, ni(t) is the additive white Gaussian noise (AWGN) with zero mean and variance, and hi is the complex channel gain of the sensing channel between the PU and the ith cognitive radio.

In cooperative spectrum sensing (CSS), each cooperative partner makes a binary decision based on its local observation and then forwards one bit of the decision Di (1 standing for the presence of the PU, 0 for the absence of the PU) to the common receiver through an error-free channel. The structure of centralized cooperative spectrum sensing in CR networks is shown in Fig. 8.

The general process is as follows: first, every CR user executes local single-node detection independently and gets detection statistic yi, second, the local dual-decision Di {0,1} is obtained by comparing yi with the detection threshold, and then, all CR users sent Di to FC; the final decision is made according to AND, M rank and OR criteria [24]. At the common receiver, all 1-bit decisions are fused together according to logic rule:

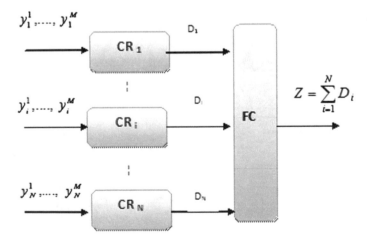

Fig. 8 Centralized cooperative spectrum sensing

$$Z = \sum_{i=1}^{N} D_i \begin{cases} \geq n, H_1 \\ < n, H_0 \end{cases} \quad (2)$$

where H0 and H1 denote the inferences drawn by the common receiver that the PU signal is not transmitted or transmitted.

For the same false alarm needs, the detection probability of CSS is higher than single-node detection, i.e. in the ideal environment cooperative sensing is better than single-node detection. On the other hand, in many cases, every CR nodes is placed in a different channel environment, so the detection performance of each user is not the same. A solution to avoid this drawback, due to the fact that even the same single node detection method is used, the detection probability of each node is not the same and consequently each CR user's detection results have different influence on the final decision is the cooperative spectrum sensing based on the weighting (CSSW) [25]. CSSW method implies screening nodes firstly from CR networks before all local decisions and weighted factors are sent to FC, and then the final information fusion is making by CR nodes obtained from nodes screening stage.

5 Cognitive Radio Testbed for WSN-UAV Collaborative Monitoring

Subsequent step in implementing the CR scheme on a network of cooperating objects composed of WSN nodes, terrestrial mobile ROVs and UAVs is a customized testbed enabling research and development on various CR schemes and algorithms. The main components of such a system are identified as: PC-based simulation environment and toolchain, emulator hardware for radio frequency network, configurable RF front-ends and digital switches [26]. In customizing the system for the particular WSN-UAV scenario, appropriate bidirectional interfaces towards relevant platforms for WSN e.g. the Contiki/COOJA environment, or UAV, e.g. the USARSim virtual development environment for autonomous robotic platforms have to be provided. Among the notable advantages on such approach, one can list flexibility, ability to perform a vast array of experiments with minimal hardware and software costs due to the reconfigurable nature of the system. Including realistic channel effects, taken from previous real measurement or predefined channel models (Gaussian, Rayleigh, Riccian, etc.). Dynamic role assignment for various topologies can also enable scalable reconfiguration among primary users, cognitive intermediate radios and common receivers. In this way, experiments become repeatable under controlled conditions and provide a first step towards deployment in real world conditions. Hybrid structure among simulated and emulated CR and SDR nodes is controlled by the experiment designer and reflects a trade-off between experiment accuracy and speed. Once the testbed results become satisfactory under dynamic environment conditions, the next step is to deploy the system under realistic conditions outside the lab.

For the initial testing scenarios, the approach described in [27] is used. This begins with a basic two user scenario in which a UAV covers a target area and communicates with the ground stations, tasked at uploading raw or aggregated data toward the mobile aerial platform. Under fundamental assumption of constant UAV altitude, hu, constant UAV speed, vu, and where D is the distance among nodes, three possible scenarios have been highlighted for modeling communication performance [27]:

- $D \gg hu$
- $D \ll hu$
- $D = O(hu)$

Among these three cases, the first one leads to low SNR values given variable path loss and leading to an alternative behaviour of the UAV above each of the users. The second scenario concludes with the UAV following a tight pattern directly above the users, minimizing path loss given similar angle-of-arrival (AoA) of the ground stations. For equalizing uplink rates the third case assumes a symmetric trajectory centered around the midpoint of the two users. By means of the CR testbed these scenarios can be further investigated and extended to the desired multi-user ground-aerial environments.

6 Conclusion

Summarizing the chapter, we have introduced and detailed a new framework and system architecture for multi-level heterogeneous monitoring and surveillance based on ground-aerial intelligent systems taking the form of wireless sensor networks (WSN) and unmanned aerial vehicles (UAVs). In order to increase the communication efficiency, a solution based on Cognitive Radio (CR) was proposed which shows promise given the specific requirements of the applications. Among their potential benefits, one can list: improving of the overall communication structure for increased availability and service level across the heterogeneous communication interfaces, reducing the effect of interferences and other link losses, better utilization of the existing spectrum and exploitation of the existing spectrum in a flexible way, increased security by mitigating for example denial of service (DoS) attacks and other forms of attacks.

The most prominent commercial application of a system implementation is linked to the automatic ground-aerial surveillance of pipeline infrastructures. Thus, by combining ground intrusion detection in a protected area along with event detection e.g. pipeline breach, with aerial monitoring by image collection and processing. At a higher decision level, data from all sources is fused for threat detection and implementing a decision support system for human operators which ultimately classify the situation and are able to dispatch intervention teams effectively in the concerned area.

Prospective work covers a pilot system implementation in a mixed simulated, emulated and real environment for subsystem level validation of the approached highlighted by this paper.

Acknowledgment The work of Grigore Stamatescu has been funded by the Sectoral Operational Programme Human Resources Development 2007–2013 of the Ministry of European Funds through the Financial Agreement POSDRU/159/1.5/S/134398. The work of Dan Popescu was supported by a grant of the Romanian Space Agency, "Multisensory Robotic System for Aerial Monitoring of Critical Infrastructure Systems" (MUROS), project number 71/2013.

References

1. Marron, P.J.: Keynote: are we ready to go large-scale? Challenges in the deployment and maintenance of heterogeneous networks of Cooperating Objects. In: 2012 IEEE International Conference on Pervasive Computing and Communications Workshops (PERCOM Workshops), p. 50, 19–23 March 2012
2. Augusto, J.C., Callaghan, V., Cook, D., Karneas, A., Satoh, I.: Intelligent environments: a manifesto. Human-Centric Comput. Inf. Sci. **3**, 12 (2013)
3. Sun, Z., Wang, P., Vuran, M., Al-Rodhaan, M., Al-Dhelaan A., Akyildiz, I.: BorderSense: border patrol through advanced wireless sensor networks. Ad Hoc Netw. **9**(3), 468–477 (2011)
4. Dorling, K., Messier, G.G., Magierowski, S., Valentin, S.: Improving aerially deployed sensor networks using cooperative communications. In: 2012 IEEE International Conference on Communications (ICC), pp. 376, 380, 10–15 June 2012
5. Stamatescu, G., Sgarciu, V.: Evaluation of wireless sensor network monitoring for indoor spaces. In: 2012 International Symposium on Instrumentation & Measurement, Sensor Network and Automation (IMSNA), vol. 1, pp. 107, 111, 25–28 Aug 2012
6. Cheng, C., Hsiao, P., Kung, H., Vlah, D.: Maximizing throughput of UAV-relaying networks with the load-carry-and deliver paradigm. In: Proceedings of IEEE WCNC 2007, pp. 4417–4424, March 2007
7. Han, Z., Swindlehurst, A.L., Liu, K.J.R.: Optimization of MANET connectivity via smart deployment/movement of unmanned air vehicles. IEEE Trans. Veh. Technol. **58**(7), 3533–3546 (2009)
8. de Freitas, E.P., Heimfarth, T., Netto, I.F., Lino, C.E., Pereira, C.E., Ferreira, A.M., Wagner, F.R., Larsson, T.: UAV relay network to support WSN connectivity. In: Proceedings of IEEE ICUMT 2010, pp. 309–314, Oct 2010
9. Palat, R., Annamalau, A., Reed, J.: Cooperative relaying for ad-hoc ground networks using swarm UAVs. In: Proceedings of IEEE MILCOM 2005, pp. 1588–1594, Oct 2005
10. Zhan, P., Swindlehurst, A.L.: Optimization of UAV Heading for the Ground-to-Air Uplink. IEEE J. Selected Areas Commun. **30**(5), 993–1005 (2012)
11. Conte, G., Hempel, M., Rudol, P., Lundstrom, D., Duranti, S., Wzorek, M., Doherty, P.: High accuracy ground target geolocation using autonomous micro aerial vehicle platforms. In: Proceedings of the AIAA Guidance, Navigation, and Control Conference and Exhibit, 2008
12. Chen, Z.L., Wei, N.Q.: Cognitive radio as enabling technology for dynamic spectrum access. Appl. Mech. Mater. **2560**(347–350), 1716–1719 (2013)
13. SDR forum page. Available on-line at http://www.sdrforum.org/pages/documentLibrary/documents/SDRF-06-R-0011-V1_0_0.pdf

14. Mitola III, J.: Cognitive Radio Architecture: The Engineering Foundations of Radio XML. Wiley, New York (2006)
15. Stamatescu, G., Popescu, D., Dobrescu, R.: Cognitive radio as solution for ground-aerial surveillance through WSN and UAV infrastructure. In: 2014 6th International Conference on Electronics, Computers and Artificial Intelligence (ECAI), pp. 51–56, 23–25 Oct 2014
16. Akan, O.B., Karli, O., Ergul, O.: Cognitive radio sensor networks. Network, IEEE **23**(4), 34–40 (2009)
17. Joshi, G.P., Nam, S.Y., Kim, S.W.: Cognitive Radio Wireless Sensor Networks: Applications, Challenges and Research Trends. *Sensors (Basel, Switzerland)*, *13*(9), 11196–11228 (2013). http://doi.org/10.3390/s130911196
18. Bicen, A., Akan, O.: Cognitive radio sensor networks in industrial applications. In: Industrial Wireless Sensor Networks, vol. 20132544. CRC Press, Boca Raton (2013)
19. Fahim, A.: Wideband spectrum sensing techniques. In: Radio Frequency Integrated Circuit Design for Cognitive Radio Systems, pp. 79–97. Springer, Berlin (2015)
20. MUROS Project, Romanian Space Agency Competition C2, 2013–2016. Available on-line: http://imtt.pub.ro/MUROS3/muros
21. Chen, H.-C., Kung, H.T., Vlah, D., Hague, D., Muccio, M., Poland, B.: Collaborative Compressive Spectrum Sensing in a UAV Environment, Paper 5.1, MILCOM (2011)
22. Patil, K., Prasad, R., Skouby, K.: A survey of worldwide spectrum occupancy measurement campaigns for cognitive radio. In: Proceeding of the International Conference on Devices and Communications (ICDeCom), pp. 1–5, Feb 2011
23. Akyildiz, I.F., Lo, B.F., Balakrishnan, R.: Cooperative spectrum sensing in cognitive radio networks: a survey. Phys. Commun. **4**, 40–62 (2011)
24. Li, Y., Shen, S., Wang, Q.: Study and optimization of cooperative spectrum sensing in OFDM cognitive radio networks. Int. J. Comput. Sci. Issues **9**(6, 3), 172–176 (2012)
25. Li, Y., Huang, H., Ye, F.: An improved cooperative spectrum sensing in cognitive radio. J. Comput. Inf. Syst. **8**(4), 1399–1406 (2012)
26. Ding, L., Sagduyu, Y., Yackoscki, J., Azimi-Sadjadi, B., Li, J., Levy, R., Melodia, T.: High fidelity wireless network evaluation for heterogenous cognitive radio networks. In: Proceedings of SPIE 8385, Sensors and Systems for Space Applications V, 83850R, May 2012
27. Feng, J., Swindlehurst, A.L.: Optimization of UAV Heading for the Ground-to-Air Uplink. IEEE J. Selected Areas Commun **30**(5), 993, 1005 (2012)
28. Yilmaz, H.B.: Cooperative spectrum sensing and radio environment map construction in cognitive radio networks. PhD thesis, Bogazici University (2012)

Using Modeling and Dynamic Simulation Techniques for Systems' Safety and Security

Gabriel Rădulescu, Emil Pricop, Marilena Nicolae
and Cosmina Roşca

Abstract This work addresses an important problem that can appear in operating the reactive distillation processes: the presence of a three-phase regime (vapor-liquid-liquid) on the column stages due to a liquid-liquid phase splitting between aqueous and organic phases. Being unwatched until several years ago, the appearance of this second liquid phase changes, at present, the way the process engineers take the reactive distillation process into account. That is to say, detecting the three phase regime can not only lead to different operating and design strategies for reactive distillation columns, but also can be an interesting actor in the field of process safety. The authors present in detail their original modeling approach which detects the existence of potential phase splitting in reactive distillation columns. Its hybrid software implementation is composed from a classical model (pseudo-homogeneous) in connection with a robust phase splitting algorithm (based on the homotopy-continuation method), performed at each simulation step. By using the proposed model instead of the classical one (which does not take into account the potential liquid-liquid phases separation), important improvements in terms of simulated system behavior, results accuracy and general safety features are obtained.

G. Rădulescu (✉) · E. Pricop · C. Roşca
Department of Automatic Control, Computers and Electronics, Petroleum-Gas University
of Ploieşti, 39 Bucuresti Blvd., 100680 Ploieşti, Romania
e-mail: gabriel.radulescu@upg-ploiesti.ro

E. Pricop
e-mail: emil.pricop@upg-ploiesti.ro

C. Roşca
e-mail: cosmina.rosca@upg-ploiesti.ro

M. Nicolae
Petroleum Processing Engineering and Environmental Protection Department,
Petroleum-Gas University of Ploieşti, 39 Bucuresti Blvd., 100680 Ploieşti, Romania
e-mail: mnicolae@upg-ploiesti.ro

© Springer International Publishing Switzerland 2016   57
E. Pricop and G. Stamatescu (eds.), *Recent Advances in Systems Safety and Security*,
Studies in Systems, Decision and Control 62, DOI 10.1007/978-3-319-32525-5_4

Nomenclature

HOLD	Molar liquid holdup on tray
J	Jacobian matrix
NC	Number of components
R	Reaction ratio
T	Temperature
V	Volumetric liquid holdup on tray
f	Function vector to be solved to 0
fgab	Vapor sidedraw molar flowrate
fgzu	External vapor feed molar flowrate
flab	Liquid sidedraw molar flowrate
flzu	External liquid feed molar flowrate
liq	Internal liquid molar flowrate
p	Pressure
psp	Saturation pressure in the vapor phase
vap	Internal vapor molar flowrate
x	Mole fraction, liquid (global)
x1	Mole fraction, liquid (phase 1)
x2	Mole fraction, liquid (phase 2)
y	Mole fraction, vapor phase
zflzu	Mole fraction in external liquid feed
zfgzu	Mole fraction in external vapor feed

Greek Letters

Φ	Phases ratio
$\gamma 1$	Activity coefficient (phase 1)
$\gamma 2$	Activity coefficient (phase 2)
η	Tray efficiency
θ	Solution vector
λ	Continuation parameter
v	Stoichiometric coefficient

Superscripts

CRIT	Critical point of the miscibility gap
PSA	Value given by the Phase Splitting Algorithm
START	Reference state (starting point for continuation)

Subscripts

A, B, C	Example states in the phase diagram
k	Tray number
i	Component indices
m	Variable indices (in the solution vector)
s	Current step

1 Introduction

Although the chemical industry age and maturity, although it is one of the most advanced controlled activity on Earth, it continuously represents a source for unexpected events, equally involving people, plants and environment. When such a failure occurs, people are practically surrounded by the chemical substances, steam and fire, all having serious injury or death. The accident may involve a large chemical cloud that spread well beyond the refinery edges, causing for people in the surrounding communities to seek medical attention during and immediately following the incident. Later on, the implications on the environment are very hard to put in equation, due to numerous and intricate inter-influences, huge costs and the required long settling time.

All over the world, multiple agencies (public, private and internal ones) opened investigations in response to such incidents, all identifying serious concerns about process safety management procedures at the refinery and expressed the need for stronger preventative safeguards. Also, these accidents provided an opportunity to take a more comprehensive look at a global industry performance. While chemical plants are subject to regulation by multiple agencies and some have developed extensive health and safety programs, additional measures and alternative approaches offer the potential for enhanced prevention and risk reduction, without imposing significant new regulatory burdens, as it is shown in [1].

But, on the other hand, such "unexpected events" don't mean only "the catastrophic ones", with obvious consequences for people and plants. Just consider another example, where a chemical plant is not correctly operated for months or maybe years.

Systems safety is a concept that means applying technologies and methods in order to identify, prevent and control hazardous situation that may occur in the functioning of the system. Safety does not apply only to operation, but it should be a critical item from the system design phase [2].

The petrochemical industry is a challenging domain in systems safety. The approach of safety has more objectives than simply avoiding accidents. When controlling a petrochemical facility, the owner should state as an objective loss prevention—financial loss, product loss and performance loss [3]. Due to the development of intrinsic safety design, that kind of hazard is more likely to produce than the traditional ones: fire or explosion. If fire hazard risk can be determined by analyzing the fire sources, fire protection measures, the risk of wrong operation in a chemical facility is very difficult to be quantified. Dynamic simulation of the process is the only way the system operator can estimate and observe the results of his decisions and operations without affecting the system.

Also, these considerations apply not only to a whole refinery or big plant, but to all actors involved in chemical industry. In this chapter we will take as example the acetic acid (AcH), widely used as a solvent or as a reagent in many chemical or petrochemical processes. Some of these processes, such as manufacturing of terephthalic acid, cellulose esters or dimethyl terephthalate, produce dilute AcH streams in compositions ranging from 1–65 wt% [4–7]. It is necessary to recover this acid and to produce aqueous stream free of organic contaminants to be discarded to the environment. This can be achieved by separating acid and water conventionally by using simple distillation or extraction, or in a modified way by using azeotropic distillation or extractive distillation. Though a simple distillation column can produce high purity acid and water (because of the absence of an azeotrope), this operation is uneconomic. Extraction cannot produce high purity acid (at one column's end) and water streams (at the other end) because of the limited phase separation and component distributions. The more customary approach therefore is to use an extra entrainer or solvent in an azeotropic distillation column or extractive distillation column, which can save up to 45 % of estimated total costs of simple distillation operation [5–7].

Another interesting approach is to react away acetic acid from a dilute AcH stream in a reactive distillation (RD) column with a suitable alcohol to produce valuable acetate. In recent years, esterification of this pure acetic acid stream with butanol in a continuous RD column has been investigated extensively. Many of these studies could achieve conversions greater than 98 % either theoretically or experimentally. On the other hand, esterification of dilute AcH stream with BuOH, for waste water treatment, is not yet fully understood. The above two processes are significantly different in a physical sense. The difference comes from the fact that in the latter process, a significant amount of water enters the column with the feed, which is not only a product of the esterification reaction but also possess the potential to form immiscible liquid phases with the other reacting components. Therefore, when the composition on a given column stage falls inside the liquid-liquid phase boundary, a liquid-liquid phase split occurs on that stage [5–7].

It is not easy to design and operate such a RD column giving high conversion of AcH. The process complicacy comes not only from the reaction—separation interactions, but also from the presence of a three phase regime (liquid-liquid-vapor). The detection of single or double liquid phases on every single tray is a crucial step for the success of a RD column design and operation. This is done by combining the main column model (MESH equations) with an extra routine for the phase split calculation [8]. The liquid-liquid phase splitting routine, used in this work, has been developed by Steyer et al. [9] based on the vapor liquid-liquid flash algorithm proposed by Bausa and Marquardt [10]. It is also used for dynamic simulation of the different column designs to study operability and sensitivity to disturbances in a second step.

2 Reactive Distillation—The Mathematical Model Summary

As previously known, the RD process integrates chemical reaction and separation by distillation in a single processing unit, having a big economical advantage over the conventional process designs, where reaction and separation are carried out in different processing units [11–13].

However, due to the strong interaction between reaction and separation, RD processes can sometimes show an intricate nonlinear dynamic behavior. Phenomena such as steady state multiplicities, self sustained nonlinear oscillations and bi-stability are only a few particular issues when dealing with the RD processes. Naturally, a profound understanding of these phenomena as well as their reliable prediction is not only of scientific interest, but also a necessary prerequisite for improved process design, industrial control and plant safety [12, 14–16].

In this work, the case of a classic RD column, as depicted in Fig. 1, is taken into account. For maximum model flexibility, the column is considered having multiple vapor/liquid feeds and/or sidedraws on trays. At top, the vapor is condensed and the resulting liquid is accumulated in decanter. Parts of top product(s) are returned as external reflux. Also, at bottom an internal reboiler is present. The reactive zone may be located anywhere inside the column, having a free catalyst load on each stage [8, 17].

A "classical" approach treats the RD process as a pseudo-homogeneous system, where no phase splitting occurs in the liquid phase [12, 18]. But, as shown above, there are cases when such a classical approach is not satisfactory. For instance, high purity products can be obtained by using a "smart" and adaptive reflux policy exploiting the miscibility gap appearance at the condenser and in the upper part of the column (a typical example being the production of butyl acetate). Also, for some

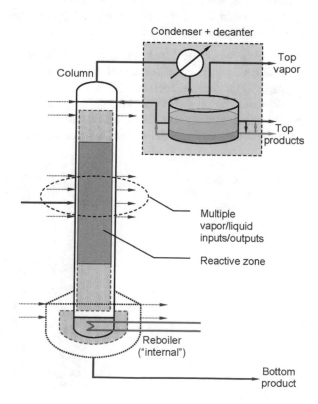

Fig. 1 Schematic representation for the RD column [8]

systems, significant differences between states in pseudo-homogeneous and heterogeneous regime can be revealed [5–7, 10, 19]. As consequence, an appropriate model has to be used in order to better reflect the real system behavior [20, 21].

2.1 The New Model Structure

As written in the open literature, the appearance of a second liquid phase makes the dynamic simulation of the reactive distillation column a much more difficult task [9, 10, 19]. The main problems are the phase state rapid, robust and reliable determination on each tray during the simulation horizon, the compositions calculation (in both phases for the trays in heterogeneous regime), phases ratio determination, managing in the same time the switches in the process model (when changes in the phase state on some trays occur).

In order to override the model switching, the authors of this work uses an original and robust approach, always considering two liquid phases; when the system leaves the heterogeneous regime, these two phases become identical [20, 21]. This way, there is no need to change the number of model equations (as some other authors revealed) when the system crosses the boundary between the homogeneous and heterogeneous region [19].

As shown in Fig. 2, a structural modeling approach was adopted in this work, considering here two sections:

- the main model, relatively close to the "classical" RD model (without phase splitting), which calculates at each step the global composition in liquid (x) and vapor (y) phases, temperature (T), internal liquid (*liq*) and vapor (*vap*) streams flowrates, for all distillation stages;
- the phase splitting algorithm, externally carried out in a separate procedure, called by the main model at each integration step, for all distillation stages; this

Fig. 2 The structural modeling approach for RD processes with potential liquid phase splitting [17]

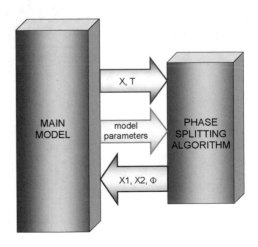

algorithm gets from the main model the global compositions (x) and temperatures (T), together with some other parameters, giving back both liquid phases compositions ($x1$ and $x2$) and ratios ($\emptyset$).

Regarding the software implementation, due to its high performances, flexibility and robustness, the author's choice is DIVA—Dynamische Simulation Verfahrenstechnischer Anlagen—working coupled with an external FORTRAN routine which run the phase splitting algorithm [22].

2.2 Modeling Principles

2.2.1 Main Model

Some basic simplifying assumptions need to be formulated, in order to have a robust and pertinent dimensional model. As the authors implemented the model in several different forms (i.e. continuous and batch distillation, homogeneously and heterogeneously catalyzed process, with or without energy balance), two assumption categories can be identified. While the particular assumptions need to be presented for each specific case, the general ones are always valid [20, 21]:

1. All column trays have constant liquid holdups.
2. The vapor holdup on trays is neglected.
3. The vapor and liquid phases are in equilibrium.
4. The reaction takes place only in liquid phases.
5. The reaction rate (R) kinetic expression is known.

In the case of kth regular tray inside the column (as shown in Fig. 3), for a homogeneously catalyzed process, with perfectly mixed reactants and catalyst, without considering the energy balance, the regular equations are [20, 21]:

Fig. 3 The kth tray inside RD column [20]

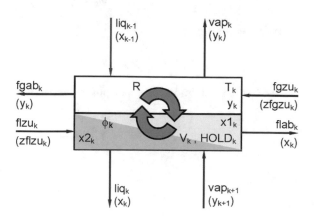

Component material balance:

$$HOLD_k \frac{dx_{i,k}}{dt} = liq_{k-1}x_{i,k-1} + vap_{k+1}y_{i,k+1} - liq_k x_{i,k} - vap_k y_{i,k}$$
$$+ flzu_k zflzu_{i,k} + fgzu_k zfgzu_{i,k} - flab_k x_{i,k} - fgab_k y_{i,k}$$
$$+ v_i \big[(1 - \phi_k) R(x1_{1,k}, \ldots, x1_{NC,k})$$
$$+ \phi_k R(x2_{1,k}, \ldots, x2_{NC,k}) \big] V_k, \quad i = 1, \ldots, NC - 1. \tag{1}$$

A global reaction rate R is considered as the linear combination between the reaction rate in phase 1 and the reaction rate in phase 2, taking into account the phases ratio. If the liquid phase splitting does not occur, then the compositions in both phases are equal and the reaction ratios are identical. The linear expression in first equation may be simplified when an equal catalyst distribution in both liquid phases is considered.

Summation condition for global liquid phase compositions:

$$\sum_{j=1}^{NC} x_{j,k} = 1. \tag{2}$$

Compositions in liquid phase 1 (externally calculated):

$$x1_{i,k} = x1_{i,k}^{PSA}, \quad i = 1, \ldots, NC \tag{3}$$

$$x2_{i,k} = x2_{i,k}^{PSA}, \quad i = 1, \ldots, NC \tag{4}$$

In this equation, $x1_{i,k}^{PSA}$ and $x2_{i,k}^{PSA}$ represents the phases composition, externally determined with the "Phase Splitting Algorithm".

Phase ratio (externally calculated):

$$\phi_k = \phi_k^{PSA} \tag{5}$$

Phase equilibrium:

$$y_{i,k}p = \eta \cdot psp_i \cdot \gamma 1_{i,k} \cdot x1_{i,k}, \quad i = 1, \ldots, NC \tag{6}$$

Summation condition for vapor phase compositions:

$$\sum_{j=1}^{NC} y_{j,k} = 1. \tag{7}$$

Total material balance for the liquid phase:

$$0 = liq_{k-1} - liq_k + flzu_k - flab_k$$
$$+ \sum_{i=1}^{NC} \left[v_i \left[(1 - \phi_k) R(x1_{1,k}, \ldots, x1_{NC,k}) + \phi_k R(x2_{1,k}, \ldots, x2_{NC,k}) \right] V_k \right]. \quad (8)$$

Total material balance for the vapor phase:

$$vap_k = vap_{k+1} \quad (9)$$

The models for the column top and bottom are also based on the equations above, with usual changes describing these slightly modified structures.

2.2.2 Phase Splitting Algorithm

As mentioned before, the phase splitting algorithm runs almost independently, checking at each step the state of all distillation stages and returning to the main model the phases compositions and ratios. It takes some mandatory information from the main model, including overall compositions, stages temperatures, coefficients for the vapor-liquid-liquid equilibrium calculation and also algorithm tuning parameters [20, 21].

The authors used in this work a phase splitting algorithm originally presented by Bausa and Marquardt [10], in the improved form proposed by Steyer et al. [9]. It is a hybrid method using a priori knowledge of phase diagram properties in order to tune-up the computational algorithm. The flash calculation is decomposed in two steps: a preprocessing step and the computational one.

In the first step, all heterogeneous regions of the system's phase diagram at the specified pressure and boiling temperature are divided into convex regions and, for each region, one reference state inside it (x_{START}, $x1_{START}$, $x2_{START}$, y_{START}, Φ_{START}, p_{START}, T_{START}), is stored—denoting here the overall composition, compositions in both liquid phases, vapor composition, phases ratio, pressure and temperature. Typically, this analyzing procedure may be carried out only once, before simulations. More, since the phase diagrams are investigated in an early phase of the process design, the information on the heterogeneous region(s) existence may be directly provided by user, at least for mixture with up to four components [20, 21].

In the next step, the difficult problem is solved by homotopy (meaning the search of non-trivial two-phase solution, $x1$, $x2$ and Φ, at some desired global composition x) by starting from a simple problem (the solution at a binary mixing gap with the composition x_{START} where the compositions of the two phases, $x1_{START}$ and $x2_{START}$, and phase ratio Φ_{START} are known). By parametrizing the overall composition with a continuation parameter λ, it can be changed from the starting composition x_{START} to the composition x, for which the phase splitting behavior has to be checked,

$$\overline{x}_i = \lambda x_i + (1 - \lambda)x_i^{START} = \overline{x}_i(\lambda), \quad i = 1, \ldots, NC \tag{10}$$

λ is changed from 0 to 1 when the continuation is performed. It can be observed that $\overline{x}_i(0) = x_i^{START}$ and $\overline{x}_i(1) = x_1$.

On its turn, the homotopy continuation algorithm is based on a repetitive two-step process. First one, the correction step, solves the following equations:

Mass balances (as constraints):

$$x1_i(1 - \phi) + x2_i\phi = \overline{x}_i(\lambda), \quad i = 1, \ldots, NC, \tag{11}$$

Activity difference equations (as necessary conditions):

$$x1_i \cdot \gamma1_i - x2_i \cdot \gamma2_i = 0, \quad 1 = 1, \ldots, NC, \tag{12}$$

The summation equation (as constraint):

$$1 - \sum_{i=1}^{NC} x1_i = 0 \quad \text{or} \quad 1 - \sum_{i=1}^{NC} x2_i = 0. \tag{13}$$

The above equations are written for the global composition $\overline{x}$ at a particular value for λ. No tray index "k" is provided, in order to increase the readability.

In the second step (predictor step), a solution θ to Eqs. (11), (12) and (13) for a new value of λ is estimated using

$$\theta_{m,s+1} = \theta_{m,s} + \Delta\theta_m = \theta_{m,s} + \frac{d\theta_m}{d\lambda}\Delta\lambda, \quad m = 1, \ldots, 2NC + 1, \tag{14}$$

θ_m denoting an element of the solution vector. For Bausa and Marquardt, θ contains $2NC$ mole fractions ($x1$ and $x2$) and one phases ratio (Φ).

The algorithm works by alternating prediction and correction steps while increasing λ from 0 to 1, effectively moving along the binodal surface in an effort to reach the desired x composition. During continuation, depending on where the studied composition x is located, three scenarios are possible, as shown in Fig. 4.

First, the test mixture with composition x_A resides in the heterogeneous region (vapor-liquid-liquid equilibrium, VLLE) and splits into the liquid phases $x1_A$ and $x2_A$. This time the continuation algorithm has to be performed only in the hetero-geneous region.

The second scenario refers to test mixture x_B, located in the homogeneous regime (vapor-liquid-liquid equilibrium, VLE), but still located below the tangent to the critical point x^{CRIT} of the miscibility gap. The first part of the continuation algorithm is performed in the heterogeneous regime, while the last part crosses the

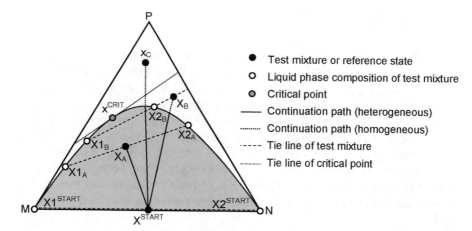

Fig. 4 Example of phase diagram for a ternary (M–N–P) system [17]

homogeneous region. Even in this case it is still possible to find a solution $x1_B$ and $x2_B$, but the phase ratio Φ is greater than 1 (which has no physical meaning).

The last case corresponds to the test mixture x_C, located in the homogeneous region, above the tangent to the critical point. In this case, no valid solution $x1_C$ and $x2_C$ is found by continuation and, like in the previous case, the algorithm says that no phase splitting occurs [8, 17].

As Bausa and Marquardt show in their paper [10], this approach is very successful in finding the correct solution very quickly, with a high reliability. However, in their original implementation the solution vector θ has $2NC + 1$ components, although the system degree of freedom is NC. This is why a modified method was used, it being developed by Steyer, Flockerzi and Sundmacher. The method's principle is to parametrize the solution vector θ by introducing so-called phase partitioning coefficients, reducing the system order to NC, as the quoted authors proved in their work [9].

The correction step is based on Newton-type iteration, where the following equation system has to be solved:

$$\theta_{s+1} = \theta_s - J^{-1}(f(\theta_s)) \cdot f(\theta_s). \tag{15}$$

In this equation, J denotes the Jacobian matrix of the remaining equation system (after model reduction), denoted here as f. To avoid inverting the Jacobian matrix, the equivalent linear equation system has to be solved. Also, for a fast and reliable solution, the authors suggest that Jacobian should be computed analytically since the equation system is highly non-linear due to the activity coefficient model used to calculate γ_i [9].

3 Model Validation

The authors decided to validate their model by reproducing the results of Brüggemann et al. [19]. Their study is focused on batch distillation process simulation in heterogeneous regime, taking as example the laboratory column for butanol esterification to butyl acetate, previously presented by Venimadhavan et al. [23].

Taking into account the existence of several low-boiling azeotropes, in such a column the butyl acetate cannot be directly recovered as the product. However, it is shown that high-purity butyl acetate could be obtained by using a clever reflux policy that exploits the appearance of a miscibility gap at the condenser and in the upper part of the column. Studying the column behavior, the quoted authors followed three operating strategies, for which they present an important amount of results:

(a) ternary non-reactive distillation (loading the column still pot with a mixture of 40 % water, 20 % butanol and 40 % butyl acetate, with no catalyst load), at a constant reflux ratio (0.9);

(b) reactive distillation (filling the still pot with a binary mixture of 51 % butanol and 49 % acetic acid), homogeneously catalyzed with sulfuric acid, at a constant reflux ratio (0.9);

(c) reactive distillation (filling the still pot with a binary mixture of 51 % butanol and 49 % acetic acid), homogeneously catalyzed with sulfuric acid, at a variable-adaptive reflux ratio (0.9 and 0.99).

The model presented in this work was adapted for a 33 stage batch column. The holdup on each tray is 0.001 kmol, the combined holdup of the condenser and decanter is 0.01 kmol and the initial holdup of the still pot is 2 kmol. Also, a constant vapor flowrate of 2 kmol/h from the reboiler is considered. The thermodynamic data (the UNIQUAC model is used for γi calculation) were taken from DECHEMA database, as a consistency condition with Brüggemann's work—even if the binary interaction coefficients from there do not correctly describe the mixture behavior [19]. As specific assumptions, no energy balance is considered (assuming a constant vapor flow from stage to stage) and a uniform liquid catalyst distribution between liquid phases has to be taken into account [8].

Figure 5 presents the schematic representation of the considered batch distillation column. The simulation results for the first two scenarios are further depicted in Figs. 6, 7, 8 and 9.

After a close analysis, the authors concluded there is a very good agreement between our diagrams and those from the original Brüggemann's paper, both for decanter and still pot, not only qualitatively-quantitatively, but also as timing, so the modeling approach in this work may be considered as valid and has to be put into value, being tested for some other applications.

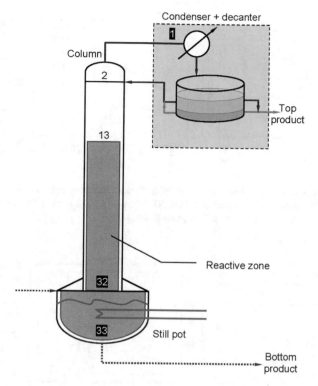

Fig. 5 Batch RD column used for validation [8]

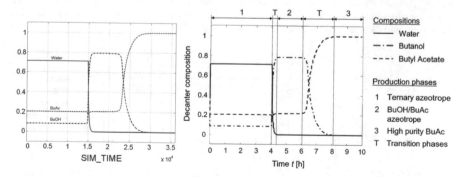

Fig. 6 Comparison between our model results (*left*) and literature results (*right*), for the first scenario—global composition in decanter. SIM_TIME is expressed in (s × 10^4) and t in (h). The *right hand side* picture is taken from Brüggemann et al. [19]

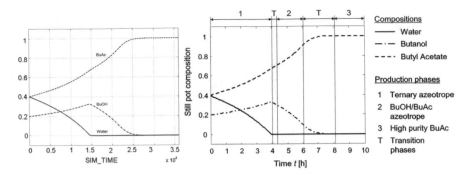

Fig. 7 Comparison between our model results (*left*) and literature results (*right*), for the first scenario—global composition in still pot. SIM_TIME is expressed in (s × 10⁴) and t in (h). The *right hand side* picture is taken from Brüggemann et al. [19]

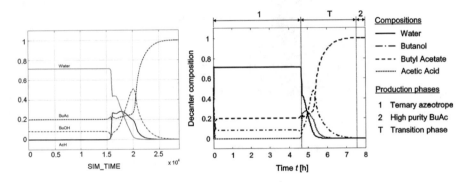

Fig. 8 Comparison between our model results (*left*) and literature results (*right*), for the second scenario—global composition in decanter. SIM_TIME is expressed in (s × 10⁴) and t in (h). The *right hand side* picture is taken from Brüggemann et al. [19]

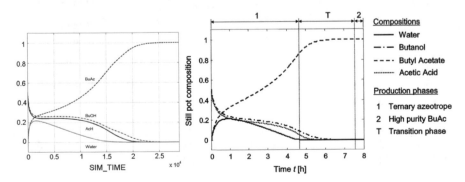

Fig. 9 Comparison between our model results (*left*) and literature results (*right*), for the second scenario—global composition in still pot. SIM_TIME is expressed in (s × 10⁴) and t in (h). The *right hand side* picture is taken from Brüggemann et al. [19]

4   A Short Case-Study: Results

At present, many studies have their focus on waste utilities treatment, especially for those associated with industrial plants. Significant emphasis is put on the recovery of dilute acetic acid from water, due to the inherent process difficulty and high environmental impact [8, 17].

The water treatment process in order to remove the acetic acid is known to be very expensive. The disposal of wastewater containing acetic acid is an important issue for the chemical industry. In order to remove this pollutant, there were developed various methods for wastewater treatment. Most of the methods, such as reverse osmosis, liquid extraction or membrane electro dialysis are not very economic efficient. Adsorption is one of the most efficient methods, but since the adsorbent used is activated carbon it keeps a high-price.

As shown before, the acetic acid cannot be easily separated from water by conventional distillation or extraction. As consequence, alternative techniques were found, one of these being the reactive distillation. In this last case, the acetic acid recovery is done through esterification (with n-butanol, for example), where a value-added ester (butyl acetate) is formed and—if the process is carefully operated—almost pure water can be withdrawn.

Figure 10 shows a 22-trays RD column, an original alternative designs proposed by the authors of this paper. At top the organic phase is totally refluxed, while the

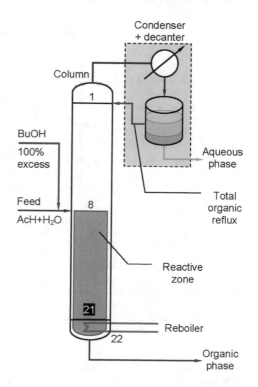

Fig. 10 RD process for acetic acid recovery from waste water

aqueous phase is withdrawn, and so the organic phase constitutes the bottom product.

The column is fed with unpurified water and excess of butanol (such as the mole ratio AcH:BuOH is 1:2) on the 8th tray, right above the reactive zone. The total feed flowrate is 0.00675 kmol/h, liquid holdup is 2×10^{-4} m^3 (per stage); the catalyst has a load of 0.0024 kg on each tray in the reactive zone.

Accordingly, the mathematical model was configured taking into account some new specific assumptions:

1. The liquid holdup on column bottom is constant.
2. The energy balance is taken into account.
3. The process is heterogeneously catalyzed.

For this column, an acetic acid conversion of 99 % was achieved, which represent a big improvement if one makes a comparison with other results announced in literature so far [20, 21].

By using the proposed modeling approach, with potential phase splitting calculation, very useful information about this configuration was obtained. The column operates in 3-phase regime at decanter level and also around the feed tray, leading to different composition profiles when the process is simulated with classical model (pseudo-homogeneous approach) and the new one (including phase stability test). As example, Figs. 11 and 12 shows the butyl acetate and water concentration along the column (in both cases).

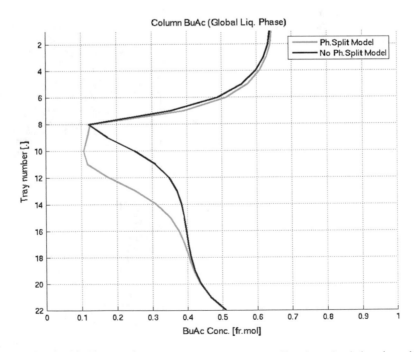

Fig. 11 Comparison between butyl acetate concentration profile when simulating the column with "phase split" model (*light gray line*) versus "no phase split" model (*dark gray line*)

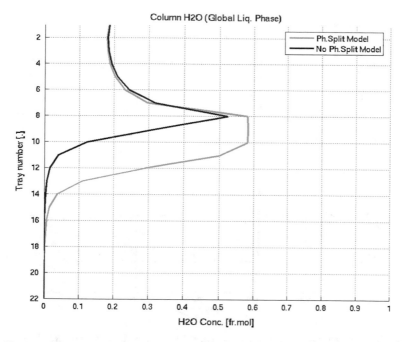

Fig. 12 Comparison between water concentration profile when simulating the column with "phase split" model (*light gray line*) versus "no phase split" model (*dark gray line*)

As remark, it can be seen that a severe drift between concentration profiles is present in the reactive zone, exactly located in the region situated in heterogeneous (liquid-liquid splitting) regime. Obviously, in this particular case one can say that by using the classical modeling approach the results accuracy is seriously affected, while the new model gives a better image on process intimacy, leading to more precise results [20, 21].

Dynamic simulation tests revealed also a very interesting feature of this configuration: a high sensitivity to disturbances (especially in feed flowrate and composition), due to traveling wave phenomena [24]. For instance, a 5 % only increase in feed flowrate (deviation from the nominal operating point) leads to a serious drop in acetic acid conversion (from 99 to 38 %, see Fig. 13), while the system moves toward a new steady state with totally different composition profiles in the reactive zone. As it can be seen in Figs. 14 and 15, the non-reactive zone above the feed tray remains unaffected. Also, Fig. 16 shows how the 3-phase regime extends from a small region around the feed tray to about 75 % of the reactive zone, without any effect in the upper part of the column.

All other simulation scenarios—not included here—confirmed the robustness and reliability of this modeling approach. More, by including the phase splitting calculation, the model describes in an improved manner the real system behavior, comparing it with the classical pseudo-homogeneous approach results.

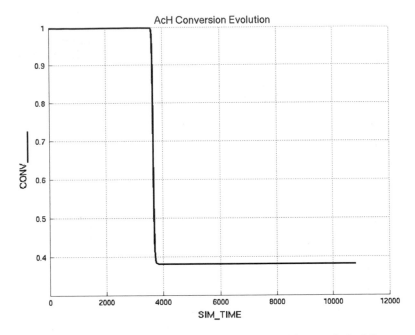

Fig. 13 The evolution of acetic acid conversion, subject to a 5 % increase in feed flowrate, after 3600 s. since the simulation start. SIM_TIME is expressed in (s)

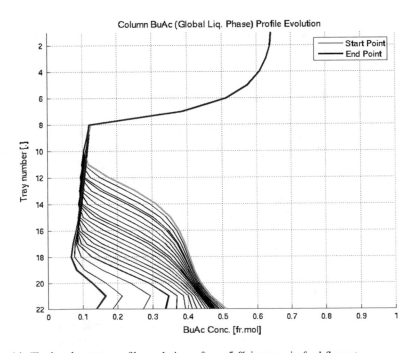

Fig. 14 The butyl acetate profile evolution, after a 5 % increase in feed flowrate

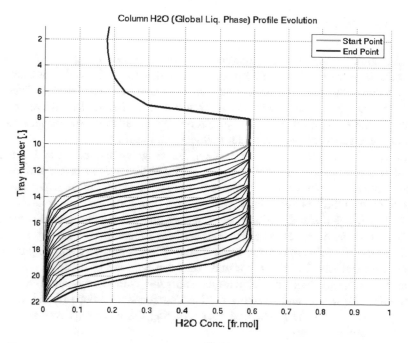

Fig. 15 The water profile evolution after a 5 % increase in feed flowrate

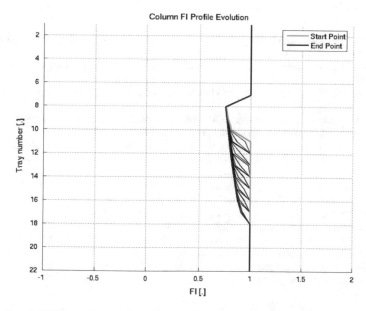

Fig. 16 The Φ ("FI") profile evolution after a 5 % increase in feed flowrate

5 Conclusion

Safety should be the most important concern of any process engineer operating a chemical or petrochemical facility. Various hazards such as fire and explosions can happen due to the failure of installation components or due to bad design. This kind of event can cause deaths, pollution and big economic losses. Luckily these accidents are not so frequent as wrong operation of an installation, another hazard that can produce substantial and constant pollution and profit loss.

In this chapter the authors focused on the usage of modelling and dynamic simulation of a reactive distillation process. The objective is to detect and avoid a very dangerous situation that can occur in the operation of this process: the apparition of a three-phase regime (vapor-liquid-liquid) on the column stages, due to the liquid-liquid phase splitting between aqueous and organic phases. Dynamic simulation is the only method to evaluate various situations without any risk. The dynamic model is able to respond to the "what if" scenarios, also showing the time when a critical situation can occur and what can be done in order to prevent its apparition. The dynamic model permits analyzing various operation scenarios, comparing the results and selecting the best approach in order to reach the efficiency objectives of the process without affecting the humans, environment and facility safety and security.

References

1. Brown, E.G.: Improving Public and Worker Safety at Oil Refineries. State of California, USA (2014)
2. Roland, H.E., Moriarty, B.: System Safety Engineering and Management. Wiley, New York (1990)
3. Leveson, N.: White Paper on Approaches to Safety Engineering (2003). http://sunnyday.mit.edu/caib/concepts.pdf
4. Saha, B., Chopade, S.P., Mahajani, S.M.: Recovery of dilute acetic acid through esterification in a reactive distillation column. Catal. Today 60, 147–157 (2000)
5. Gangadwala, J., Rădulescu, G., Kienle, A., Sundmacher, K.: Computer aided design of reactive distillation processes for the treatment of waste waters polluted with acetic acid. Comput. Chem. Eng. 31(11), 1535–1547 (2007)
6. Gangadwala, J., Rădulescu, G., Paraschiv, N., Kienle, A., Sundmacher, K.: Dynamics of Reactive Distillation Processes with Potential Liquid Phase Splitting. In: ELSEVIER's Computer-Aided Chemical Engineering Series, vol. 24, pp. 213–218 (2007)
7. Gangadwala, J., Rădulescu, G., Kienle, A., Steyer, F., Sundmacher, K.: New process for recovery of acetic acid from waste water. Clean Technol. Environ. Policy 10, 245–254 (2008)
8. Rădulescu, G., Gangadwala, J., Kienle, A., Steyer, F., Sundmacher, K.: Dynamic simulation of reactive distillation processes with liquid-liquid phase splitting. Buletinul Universității Petrol-Gaze din Ploiești, vol. LVIII, Seria Tehnică, Nr. 1/2006 (2006), pp. 1–12. ISSN 1224-8495
9. Steyer, F., Flockerzi, D., Sundmacher, K.: Equilibrium and rate-based approaches to liquid-liquid phase splitting calculations. Comput. Chem. Eng. 30, 277–284 (2005)

10. Bausa, J., Marquardt, W.: Quick and reliable phase stability test in VLLE flash calculations by homotopy continuation. Comput. Chem. Eng. **24**, 2447–2456 (2000)
11. Gangadwala, J., Mankar, S., Mahajani, S., Kienle, A., Stein, E.: Esterification of acetic acid in the presence of ion-exchange resins as catalysts. Ind. Eng. Chem. Res. **42**, 2146–2155 (2003)
12. Sundmacher, K., Kienle, A.: Reactive Distillation—Status and Future Directions. Wiley-VCH, Weinheim (2002)
13. Singh, A., Hiwale, R., Mahajani, S.M., Gudi, R.D., Gangadwala, J., Kienle, A.: Production of butyl acetate by catalytic distillation. Theoretical and experimental studies. Ind. Eng. Chem. Res. **44**, 3042–3052 (2005)
14. Gangadwala, J., Kienle, A., Stein, E., Mahajani, S.: Production of butyl acetate by catalytic distillation: process design studies. Ind. Eng. Chem. Res. **43**, 136–143 (2004)
15. Grüner, S., Mohl, K.-D., Kienle, A., Gilles, E.D., Fernholz, G., Friedrich, M.: Nonlinear control of a reactive distillation column. Control Eng. Pract. **11**, 915–925 (2003)
16. Luyben, W.L., Pszalgowski, K.M., Schaefer, M.R., Siddons, C.: Design and control of conventional reactive distillation processes for the production of butyl acetate. Ind. Eng. Chem. Res. **43**, 8014–8025 (2004)
17. Rădulescu, G., Gangadwala, J., Paraschiv, N., Kienle, A., Sundmacher, K.: Dynamics of reactive distillation processes with potential liquid phase splitting based on equilibrium stage models. Comput. Chem. Eng. **33**(3), 590–597 (2009)
18. Taylor, R., Krishna, R.: Modeling reactive distillation. Chem. Eng. Sci. **55**, 5183–5229 (2000)
19. Brüggemann, S., Oldenburg, J., Zhang, P., Marquardt, W.: Robust dynamic simulation of three-phase reactive batch distillation columns. Ind. Eng. Chem. Res. **43**, 3672–3684 (2004)
20. Rădulescu, G., Paraschiv, N., Mihalache, S.: A systematic approach on the dynamic modeling of reactive distillation processes. The standard mathematical model (NPHSP). Rev. Chim. **64** (9/2013), 1043–1046 (2013)
21. Rădulescu, G., Mihalache, S.F., Popescu, M.: A systematic approach on the dynamic modelling of reactive distillation processes with potential liquid phase splitting. Building-up the improved PHSP simulation model. II. Revista de Chimie **65**(6), 718–724 (2014)
22. Mangold, M., Kienle, A., Mohl, K.D., Gilles, E.D.: Nonlinear computation using DIVA—Methods and applications. Chem. Eng. Sci. **55**, 441–454 (2000)
23. Venimadhavan, G., Malone, M.F., Doherty, M.F.: A novel distillate policy for batch reactive distillation with application to the production of butyl acetate. Ind. Eng. Chem. Res. **38**, 714–722 (1999)
24. Grüner, S., Kienle, A.: Equilibrium theory and nonlinear waves for reactive distillation columns and chromatographic reactors. Chem. Eng. Sci. **59**, 901–918 (2004)

HAZOP-Based Security Analysis for Embedded Systems: Case Study of Open Source Immobilizer Protocol Stack

Jingxuan Wei, Yutaka Matsubara and Hiroaki Takada

Abstract Nowadays, with the introduction of network connectivity both inside and outside modern vehicles, researchers have identified that the system is actually fragile if an attacker could locate any security vulnerabilities of the system. Although security analysis techniques prospered in the industry, still a general, compatible, and effective one remains uncertain. This chapter aims to transplant the safety analysis technique HAZard and OPerability studies (HAZOP) into an appropriate security analysis technique. By conducting a case study of security analysis for Open Source Immobilizer Protocol Stack, we demonstrate the usability of the proposed technique and discusses results of the analysis.

1 Introduction

In recent years, integration of Electronic Control Units (ECUs) in all kinds of machinery has brought us great convenience as well as a remarkable increase of production and operation efficiency. Automobiles, for example, have taken great advantages of ECUs, such as instructing drivers to drive safely and comfortably, as well as assisting technicians to conduct proper diagnosis when the vehicle is under maintenance. These digital components oversee a broad range of functionalities such as the drive train, brakes, lighting and entertainments, etc. In fact, few operations are not controlled by the computers embedded in a modern vehicle. It is said that a modern luxury automobile contains up to 70 different ECUs within millions of lines of software code [1]. In order to cooperate all the ECUs, embedded system software plays a significant role. From the beginning of the vehicles design, no

J. Wei (✉) · Y. Matsubara · H. Takada
Nagoya University, Furo-Cho, Chikusa-Ku, Nagoya, Aichi, Japan
e-mail: jx@ertl.jp

Y. Matsubara
e-mail: yutaka@ertl.jp

H. Takada
e-mail: hiro@ertl.jp

© Springer International Publishing Switzerland 2016
E. Pricop and G. Stamatescu (eds.), *Recent Advances in Systems Safety and Security*,
Studies in Systems, Decision and Control 62, DOI 10.1007/978-3-319-32525-5_5

matter it is speeding on the road, providing entertainment contents to passengers, or even communicating with outer network to upgrade the software, the embedded system never be absent and is always running in the background across the entire life cycle of an automobile.

The complexity of safety-critical embedded systems such as areophane and automotive control systems has increased rapidly. Until recently, such kinds of industry have mainly focused on the safety design of an embedded system. In fact, although several international standards for functional safety such as IEC 61508 [2] and ISO 26262 [3] have been published, security of the safety-critical systems is not mentioned in such standards. The guarantee of every functionalities proper running is the primary task all the way from the initial design, even to the maintenance during the vehicles service. Techniques have been introduced to designers to prevent every piece of equipment in the vehicle from fatal malfunction, in order to protect passengers lives. Failure Mode and Effect Analysis (FMEA) and Fault Tree Analysis (FTA) are two general techniques that could help analyzers to consider about many of the safety concerns thoroughly and efficiently. However, with the debut of network connection availability for embedded systems, highly digitalized modern vehicles are exposed to information attacks via all kinds of unauthorized, either wired or wireless connection. It is not good enough anymore for the vehicle to just function fine as initially designed, it is also required that the vehicle possesses the ability to defend itself from attacks coming from outside connection. Security vulnerabilities have been detected among modern vehicles, if intentionally exploited by attackers, making the vehicle itself as well as passengers and their properties inside the vehicle exposed in danger [4].

Conducting a safety analysis only eliminates concerns that without peoples intention to deliberately act as villains. While considering the intentionality, despite the assurance that vehicle will be working properly by itself, there could also be a security concern that someone may be exploiting certain security vulnerability to deliberately launch attacks toward the vehicle and causing the vehicle to be compromised eventually. To address such problem, a more systematic approach of security analysis is often needed.

This chapter presents an analysis technique to help embedded system designers conducting security analysis at the phase of system design. During the analysis, threats should be eliminated as much as possible by implementing related countermeasures. Based on the fundamental idea from HAZOP, this chapter refines a new security analysis technique by changing the original guidewords into 8 actions extracted from the attack taxonomy of the taxonomy Computer Emergency Response Team (CERT) [5]. This HAZOP-based security analysis technique uses these actions as the new guidewords to examine a given system architecture and uncover the security vulnerabilities from the design.

During the security analysis, applying the Guidewords to the given system architecture will help us to find out unconsidered deviations (unexpected functionality, unwanted connection, etc.) of the system. To consider the causes and the consequences of certain deviation, this chapter discusses the deviations local effects (affecting systems normal functionality) as well as the global effects (keeping the user from properly using the system). On summarizing all the analysis result entries into one overall table, this chapter also issues a severity value to each of the result entry in order to conduct a risk evaluation and argue about whether or not further precautions should be implemented during the system design. To visualize the given system architecture, in this chapter, we use a sequence diagram to assist analyzing the security concerns.

To demonstrate the applicability of the proposed technique, this chapter describes a simple case study of Open Source Immobilizer Protocol Stack (OSIPS) developed by Atmel [6]. In the case study, this chapter analyzes security vulnerabilities within the OSIPS, to generate a comprehensive and accurate analysis results. By comparing the final analysis results to detected security vulnerabilities [7] of OSIPS, this chapter discusses effectiveness and applicability of the proposed security analysis technique.

This chapter makes the following contributions:

- A new HAZOP-based security analysis technique.
- An analysis sheet containing threats and hazards found in given system architecture.
- A case study for the OSIPS.

This chapter is organized as follows: in Sect. 2, we introduce some related works including the safety analysis techniques as well as some security analysis techniques. In Sect. 3, we present a detailed security analysis flow and the proposed security analysis technique to find threats in the system design phase. In Sect. 4, we describe the results of case study of security analysis for the OSIPS. In Sect. 5, we discuss effectiveness and applicability of the proposed method. Finally, in Sect. 6, we conclude this chapter.

2 Related Works

2.1 Safety Analysis Techniques

Before talking about security analysis, the fact could not be ignored that for many years safety analysis techniques have been able to keep the machinery from malfunction, preventing humans from getting injured or even fatal casualties. Three of main safety analysis techniques used in the industry are listed as,

FMEA, treats each component that makes up the system as the analysis object. Starting from these components, safety analysis experts are focused to list up all the failure modes that may occur in each component of the system. Then, experts will consider about a chain of all possible effects led by the particular failure mode, and at last evaluate the severity of those effects that will eventually cause accidents to either the system, or the user.

FTA, is a technique that firstly takes each ultimate fault of the whole system as the root of a tree, and then, to grow leaves on such tree, starts to consider all the causes that will ultimately lead to such fault. Presenting a tree with a root node of system failure or hazard and leaf-nodes of all kinds of faults of the system that lead to such system failure [8].

HAZOP, takes up a whole new viewpoint neither from the reasons causing any system failure nor the supposed consequences of any system faults, but takes the execution flow as the analysis object during a simulation of the systems running. Mostly used for building chemical plants, in order to uncover deviations as much as possible, a team of experts performing HAZOP technique will systematically consider each process unit in the plant, such as pipelines, tanks, or reactors, and each hazard one by one, raising queries about the design. Based on the description of the chemical plant design, experts will use the guidewords such as NO(NOT, NONE), MORE, LESS, AS WELL AS, PART OF, REVERSE and OTHER THAN to examine every variables of interest such as flow in a pipeline, temperature of a reactor, pressure of a tank, level of composition, or time of a reaction. Each guideword will be applied to every single line in the design one by one, to examine if there would be any deviations away from the original design intentions [9]. However, the guidewords are not suitable to analyze the software architecture or data flow.

2.2 Security Analysis Techniques

Now that, security problems have become a general concern when developing an embedded system, and certain techniques assisting system designers to perform a security analysis have also been introduced to the industry. Such as,

Attack tree. This technique can be used for representing attacks against a system. The nodes in the tree represent the different possible steps in the attack, with the top-node as the goal of the attack. All nodes below the top-node can be assigned quantitative or qualitative values. If the former type of values are assigned, it is possible to do various calculations of the top-node as well. A textual list structure is recommended if the attack trees become too complex. However, such analysis technique relies heavily on the related experience of the analyzer, and may not be a suitable choice for people who is still not quite familiar to security concerns.

SafSec, in which, Saf is for Safety and Sec is for Security. As displayed in the name, this is a method of managing both safety and security risks in a system development project, especially those of advanced avionics architectures (AAvA) or integrated modular avionics (IMA). Consisting of Guidance Material [10] and Standard [11], the SafSec methodology can be used to ensure that the assurance provided through safety and security certification is met efficiently with minimum rework to enable IMA to be realized and cost benefit provided [12]. However, such method focused mainly on the analysis of modularized avionics production, and the application of automobile embedded system software still remains unclear.

STPA-Sec. System-Theoretic Process Analysis for Security, is a top-down, system engineering technique modified from safety analysis technique STPA. STPA-Sec identifies security vulnerabilities and requirements as well as scenarios leading to violation of security constraints. Then uses the results to refine system concept and makes it to be more secure [13]. The hazard analysis process in STPA-Sec consists of five phases, which are (a) Determining unacceptable losses. (b) Creating a model of the high level control structure-HLCS. (c) Identifying unsafe/unsecure control actions. (d) Developing security requirements and constraints. (e) Identifying casual scenarios [14]. By conducting this 5 phases, STPA-Sec can address technical and organizational issues and supports a security-driven concept development process where vulnerability analysis influences and shapes early design decisions, also iterated and refined as concept evolves [13].

SHARD. Software Hazard Analysis and Resolution in Design, brings the fundamental idea from HAZOP to the field of software by using guide words to suggest a deviation. There are 5 guide words:

- OMISSION
- COMMISSION
- EARLY
- LATE
- VALUE

as shown in Table 1. However, as the author has mentioned, a single guide word may have many interpretations, and this could cause each interpretation suggesting different deviations. That is to say, security analysis conducted by different people could produce all kinds of different results and not unified. In our research, we want to stand on the idea of SHARD and HAZOP, but we made some revision by changing SHARD's guide words into some particular actions. More details will be introduced in later sections.

So, in general, despite the prosperity of security analysis techniques, a general, compatible, and effective technique for analyzing security vulnerabilities in automobile embedded system still remains uncertain in the industry. This chapter tries

Table 1 Guide words from SHARD [8]

Guideword	Meaning
OMISSION	The service is never delivered, i.e. there is no communication
COMMISSION	A service is delivered when not required, i.e. there is an unexpected communication
EARLY	The service (communication) occurs earlier than intended. This may be absolute (i.e. early compared to a real-time deadline) or relative (early with respect to other events or communications in the system)
LATE	The service (communication) occurs later than intended. As with early, this may be absolute or relative
VALUE	The information (data) delivered has the wrong value

to redirect the safety analysis approach of HAZOP, to the security analysis for designing automobile embedded systems. Now, based on the fundamental idea of applying guidewords from HAZOP, this chapter takes efforts to transplant this safety analysis technique into a security analysis technique.

3 Security Analysis

3.1 Security Analysis Flow in System Development

It is often considered impossible to develop a perfect embedded system with no security flaws. In fact, even if such system does exist, it would probably be high in cost but low in efficiency. However, there is a practical engineering approach suggesting to predefine the value of every asset that needs protection, and decide which asset to protect according to their values. If an attacker would have to pay much more expensive efforts but only to gain little, he would probably give up on launching such attack. Thus the asset is considered as relatively secured.

Process of developing embedded system software is much more like the development of a consumer software product. During the phase of product definition, engineers discuss the concept of operations to clarify the systems goal: what to do with this system. On top of the decision of the system concept, engineers sketch out system requirements and architecture to further discuss the system functionality. And it is this phase during which a security analysis should be conducted in order to eliminate or mitigate potential security treats as much as possible. To examine such issue, Fig. 1 shows a practical security analysis flow.

Fig. 1 Security analysis flow

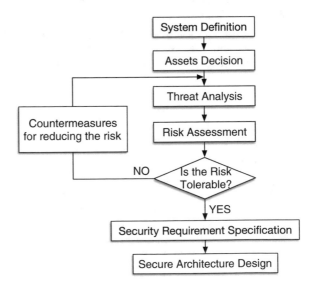

Starting from the choosing which asset to protect, security analysis will begin and further argue about threats that will cause the assets to be compromised. The fact that certain assets have been compromised will be considered as a risk, and will be assessed by experts to evaluate how hazardous this risk is going to be. At this point, qualitative calculation for a severity value as well as the occurrence rate will be issued for the evaluation, in order to argue about whether this risk is tolerable. If such risk is tolerable and could be ignored during the design, then it is safe to establish the security requirement specification concerning the current asset and continue the design. On the other hand, if such risk is not tolerable, and is considered as hazardous enough to cause serious accidents in the future affecting users life or properties, then it is necessary to consider practical countermeasures, which also need to be at an acceptable cost, for minimizing the risk. In order to make sure that whether the countermeasures are coming into force or not, threat analysis and risk assessment will be performed again until risks found during the analysis are finally tolerable.

3.2 Threat Analysis

3.2.1 Objectives

Threat analysis is conducted to locate potential security concerns in the system. Environmental threats, for example, a sudden power failure will cause the system to black out and stop functioning, or an unexpected struck by lightning causing a short inside the system. On the other hand, artificial threats are brought by humans with or without intentionality. Intentional artificial threats are brought by people who are deliberately exploiting systems vulnerability for malicious gain, such as eavesdropping on the communication, or falsification of the data during the transportation in order to send fake information. However, non-intentional artificial threats are due to negligence during the system operation, such as a misoperation or malfunction. These are all considered as potential threats in the system, and precautions should be implemented into the design before the systems roll out. So, here we just list up a few of all the threats that may be concealed in the system design at an early stage, and here is the question: how can we guarantee that the conducted threat analysis is exhaustive and comprehensive? Can we assure that all the potential threats have been pointed out and that the whole system is free from security concerns?

It may be difficult to guarantee a system that has no security flaws, however, with proper security analysis technique applied, we are still able to sweep out the threats as much as possible. And when most of the security concerns have been controlled at a minimum level, making the effort to attack such system to be at an extremely expensive level, then, it is suitable to say that the system now is under a relatively safe environment.

Table 2 Guidewords for detailed system level analysis, with their meanings

Guideword	Meaning
PROBE	Access a target in order to determine its characteristics
SCAN	Access a set of targets sequentially in order to identify which targets have a specific characteristic
FLOOD	Access a target repeatedly in order to overload the targets capacity
AUTHENTICATE	Present an identity of someone to a process and, if required, verify that identity, in order to access a target
SPOOF	Masquerade by assuming the appearance of a different entity in network communications
BYPASS	Avoid a process by using an alternative technique to access a target
MODIFY	Change the content or characteristics of a target
READ	Obtain the content of data in a storage device, or other data medium

3.2.2 Target Diagram

Diagrams that visualize the design and correctly explain all the functionality as well as data flows are necessary to perform a proper security analysis. One of the popular diagram choices is Unified Modelling Language (UML). The standard UML diagrams are: class diagram, object diagram, use case diagram, sequence diagram, state chart diagram, activity diagram, collaboration diagram, component diagram, and deployment diagram. In this chapter, we use sequence diagrams to depict sending and receiving messages among devices.

3.2.3 Guidewords

A modified attack taxonomy in [5] provides actions related to security attacks. As shown in Table 2, they are PROBE, SCAN, FLOOD, AUTHENTICATE, SPOOF, BYPASS, MODIFY, and READ. The detail definitions of the actions are found in [15]. We employ the actions as guidewords for threat analysis in the system architecture and data flow, and derive deviations from the requirements and designs of the system by applying the guidewords.

We divided the guidewords into 2 groups, Primary Guidewords and Secondary Guidewords. Primary Guidewords include actions to acquire information about the target:

- Probe
- Scan
- Read

And secondary Guidewords include actions to utilize the information acquired by Primary Guidewords and execute high-level attacks:

- Flood
- Authenticate

- Spoof
- Modify
- Bypass

A special case for Primary Guidewords is that, attacker with concrete knowledge or former experience of the target could skip attempting the actions in Primary Guidewords to investigate the target and commence attack right away with actions in Secondary Guidewords. Therefore, when conducting a detailed analysis on the message level to examine the data flow, first of all, each guideword from 2 groups will be applied to the messages to conduct the analysis. Then one guideword from Primary Guidewords and another guideword from Secondary Guidewords will be combined to apply to the messages and begin the analysis one more time.

3.2.4 Analysis Sheet

With the application of each guideword (or combination of two guidewords), entries of analysis result will be generated one by one. To summary all the entries as a single table, this chapter creates an analysis sheet as shown in Table 3. Each guideword will be applied to the data flows in turn followed by a list containing all possible attacks, and generates deviations leading to local effects and global effects. And a severity value will also be issued according to the after-defined severity table to the global effects. The combination of 2 guidewords from different group will also be applied to the analysis objects. Starting from 8 cells that represent the status (whether or not has been applied as the guideword. • as yes, and − as no) of the guideword up ahead. Here as shown in the sample is an entry of applying the combination of 2 guidewords, which are Read and Flood.

3.3 Risk Assessment

A proper risk evaluation should be conducted by considering the combination of threat severity, the threat occurrence probability, as well as the threat success probability. With the given formula (1),

$$Risk = Severity \times Occurrence\ Probability \times Threat\ Success\ Probability \quad (1)$$

Table 3 Analysis sheet: sample

Probe	Scan	Read	Flood	Authenticate	Spoof	Modify	Bypass	Possible attacks	Deviation	Local effect	Global effect
−	−	•	•	−	−	−	−	−	−	−	−

a risk value should be easily calculated to help identifying whether such risk is tolerable or not. For example, a risk value larger than 5 suggests that the risk is non-tolerable and proper countermeasures should be implemented to eliminate such risk; while a risk value less than 4 suggests that the risk is tolerable and could be ignored during the execution of the system.

4 A Case Study

After the text edit has been completed, the paper is ready for the template. Duplicate the template file by using the Save As command, and use the naming convention prescribed by your conference for the name of your paper. In this newly created file, highlight all of the contents and import your prepared text file. You are now ready to style your paper; use the scroll down window on the left of the MS Word Formatting toolbar.

4.1 Open Source Immobilizer Protocol Stack (OSIPS)

The OSIPS is developed by Atmel. This stack is intended to be used in conjunction with the automotive transponders chips of Atmel, but also in general, could be deployed in any other compatible transponders chips from other manufacturers as well. Specifications, designs and source-code of the OSIPS can be obtained as open-source system [6].

This stack comes with a variety of commands issued by the reader, and sent to the key fob. During a normal use of the immobilizer, the car will act as the reader sending commands to authenticate with the registered key fob. However, during a diagnostic or maintenance, the reader could also be used as a programming device for the car manufacturer or distributor. The communications between the car and the key fob are implemented at a low frequency of 125 kHz, which is a very limited range, practically just a few centimeters [7].

The command set out in the protocol stacks specification encompasses eleven commands. They include reading of the key fobs unique ID (UID) and error status, initiation of authentication, setting of the used secret keys, initiation and leaving of the so-called enhanced mode (for RF communication powered by the battery), a request to repeat the last response, reading and writing of user memory as well as setting memory access protection to certain memory sections. Authentication can be configured to be as [7]:

- Unilateral, only key fob authenticates itself to the reader.
- Bilateral, both key fob and reader authenticate themselves to each other.

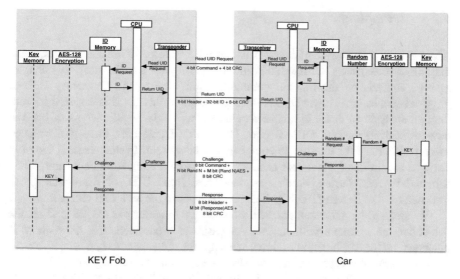

Fig. 2 Sequence diagram: unilateral authentication

This chapter uses the unilateral authentication as the analysis object for a case study of the proposed HAZOP-based security analysis technique. The authentication follows a general pattern of exchanging challenge and response information between the communicating two devices. A challenge will be sent by the verifying party to the participant party often containing a randomly generated number. Then, on receiving the challenge, the participant encrypts it with a key that is previously shared with both parties, and sends it back as the response. The verifying party compares the response to its own encryption result with the same key, if matched, then authentication succeeds.

Figure 2 shows a conceptual diagram demonstrating the unilateral authentication process of the OSIPS. Base on the specification, this chapter tries to analyze the security concerns from the messages level using the sequence diagram shown in Fig. 2. Detailed data flows will be examined one by one using the pre-defined new guidewords, in order to perform a deeper security analysis.

As described in [6], the authentication is initiated by sending a Read UID command from the car. The key fob receives such command, and read the UID information from its own memory and send it back to the car. The car will verify whether the UID is the same with the UID information in its own memory. If the UID matched, the car will start the authentication by sending a challenge mainly composed of a randomly generated number. The key fob received such challenge, read the encryption key from its own memory, and use it to encrypt the challenge as the response and sent it back. The car verify whether this response is same result as its own encryption result. If the result matched, then at last the authentication succeeds, and user may use the key fob to start operating the car.

4.2   Assumptions

Alongside with the communication between the key fob and the car, this sequence diagram also indicates the internal communication within each party. this chapter ignores analysis on the internal communications due to the temporary lack of physical connection. However, this does not implicit that there is no need to conduct an analysis on them, in fact, if an attacker was able to gain access of internal devices such as through OBD-II port, USB, iPod Dock, etc., there will be even more possibilities for him/her to compromise the system. In this chapter, we only perform security analysis on the external wireless communication between the key fob and the car, during which command like Read UID, or the exchange of information like Return UID, Challenge or Response are sent and received.

We assume that attackers are lack of knowledge about the OSIPS and all the subjective are attackers with malice. As we have mentioned, risk assessment is evaluated by the combination of threat severity, the threat occurrence probability, as well as the threat success probability. However, during the security analysis in this case study, it is still considered as difficult to calculate both probabilities of threats occurrence and success. Therefore, in this chapter, a given severity value is solely used to conduct a simplified version of risk evaluation.

4.3   Threat Analysis

As shown in from Tables 4, 5, 6 and 7, analysis results of applying the combination of Primary Guidewords and Secondary Guidewords are all summarized in the analysis sheet.

As the analysis results in the analysis sheet, deviations by the application of each guideword as well as the combination of Primary Guidewords and Secondary Guidewords are listed up to consider all the possible local effects and global effects. Comparing the possible attacks reported in [7], in which the security issues include (not limited to) as relay attack with genuine key fob, tracking, denial-of-service attacks, replay attack on authentication, spoofing attack on memory access protection, and hijacking communication sessions, we could find most of attacks. This results imply the effectiveness and applicability of the proposed method in the security analysis.

Applying all Primary Guidewords to messages lead to the same result due to same interpretation of Probe, Scan, and Read, therefore all analysis results relating to Primary Guidewords are summarized under a single Primary Guideword: Read. Also, the guidewords such as Authenticate and Bypass were not applicable for applying to the analysis objects. For example, Read UID request does not contain any specifications or data flows about authentication between the car and the key

Table 4 Analysis Sheet: Read UID

R[1]	F[2]	S[3]	M[4]	Possible attacks	Deviation	Local effect	Global effect
•	•	–	–	Denial of Service attacks	Flooding the KEY Fob with Read-UID-like request, which makes the KEY Fob unable to receive and deal with connections any more	The KEY Fob will not be able to send the UID information to the car	Failure of the exchange of UID information between a registered KEY Fob and the car. The authentication will not be triggered regardless of the users request
•	–	•	–	Tracking	Transponders can be used to relay the communications between the car and the key fob	Without the genuine key fob being in the communication range, the car will be tricked to send a Read UID request to a transponder near the car	Without users intention, the key fob will receive a Read UID request through a transponder near the key fob. Person with the KEY Fob that is associated with a specific UID could be tracked down for their whereabouts
•	–	–	•	Unauthorized falsification	Falsification of data during the transportation	A non-Read UID request will be sent to the key fob	Unauthorized falsification will be ignored by the verification of CRC checksum

R[1] Read, F[2]: Flood, S[3]: Spoof, M[4]: Modify

fob; or that a Read UID request is required to initiate the authentication between the car and the key fob, and cannot be bypassed. The same situation also appeared in conducting analysis on Return UID, Challenge and Response. Therefore, we could omit analysis related to Authenticate and Bypass.

4.4 Risk Assessment

As shown in Table 8, all global effects that have been listed up in from Tables 4, 5 and 6 are now summarized together and evaluated about their severity one by one. Noted that there is not a specific rule established on how to decide the severity value of each global effect. We believed that every company or organization has

Table 5 Analysis sheet: return UID

R[1]	F[2]	S[3]	M[4]	Possible attacks	Deviation	Local effect	Global effect
•	•	–	–	Denial of service	Flooding the car with Return-UID-like information, which makes the car unable to receive and deal with connections any more	The car will not be able to receive the UID information from KEY Fob	Failure of the exchange of UID information between a registered KEY Fob and the car. The authentication will not be triggered regardless of the users request
•	–	•	–	Relay attack	The car will receive UID information within the Return UID from Unregistered key fob or unknown device	By constantly sending Return UID until a challenge is received, the attacker may be able to acquire the UID information stored in the car	With the UID information in hands, it just extends the possibility to launch all kinds of attack
•	–	–	•	Unauthorized falsification	Falsification of data during the transportation	A non-Return UID request will be sent to the key fob	Unauthorized falsification will be ignored by the verification of CRC checksum

R[1] Read, F[2]: Flood, S[3]: Spoof, M[4]: Modify

their own rules, but this chapter considers severity value evaluation from 3 prospects: Automobile, Passengers well being and Privacy. And at last, a total severity value will be generated by calculating the average of the 3 non-zero prospects' value. Value larger than 5 will be considered as risk being non-tolerable and value less than 4 will be considered as risk being tolerable. In this case study, global effect with the ID of 1, 2, 4, 5, 6, 7 are considered as dangerous hazards and should be minimized or eliminated during the system design.

5 Discussion

As explained at the previous section, we could confirm the effectiveness and applicability of our method. However, there are still several problems that need further consideration. This combination of guidewords generates a new problem:

Table 6 Analysis sheet: challenge

R[1]	F[2]	S[3]	M[4]	Possible attacks	Deviation	Local effect	Global effect
•	•	–	–	Denial of service	Flooding the KEY Fob with challenge-like information, which makes the KEY Fob unable to receive and deal with connections any more	The KEY Fob will not be able to receive the challenge information from the car	Failure of the exchange of challenge information between a registered KEY Fob and the car. The authentication will fail regardless of the users operation
•	–	•	–	Replay attack	Attacker could pretend to be the car and send challenges to the key fob	Challenge information recorded from eavesdropping on a genuine authentication will be sent to key fob	Attacker can get himself/herself authenticated with the recorded challenge information
•	–	–	•	Unauthorized falsification	Falsification of data during the transportation	A non-Challenge request will be sent to the key fob	Unauthorized falsification will be ignored by the verification of CRC checksum

R[1] Read, F[2]: Flood, S[3]: Spoof, M[4]: Modify

Table 7 Analysis sheet: response

R[1]	F[2]	S[3]	M[4]	Possible attacks	Deviation	Local effect	Global effect
•	•	–	–	Denial of service	Flooding the car with response-like information, which makes the car unable to receive and deal with connections any more	The car will not be able to receive the response information from KEY Fob	Failure of the exchange of response information between a registered KEY Fob and the car. The authentication will fail regardless of the users operation
•	–	•	–	Not available	The car will receive a response from Unregistered key fobs or unknown devices	The car will receive a fake response	Without the genuine KEY to correctly encrypt the information, this fake response will be rejected at the car
•	–	–	•	Unauthorized falsification	Falsification of data during the transportation	A non-Response request will be sent to the key fob	Unauthorized falsification will be ignored by the verification of CRC checksum

R[1] Read, F[2]: Flood, S[3]: Spoof, M[4]: Modify

Table 8 Severity value table for risk evaluation

ID	Global effect	User awareness	A[1]	U[2]	P[3]	S[4]
1	Failure of the exchange of UID information between a registered KEY Fob and the car. The authentication will not be triggered regardless of the users request	Users will not notice any unusual change until they try to operate the car	8	0	0	8
2	Without users intention, the key fob will receive a Read UID request through a transponder near the key fob. Person with the KEY Fob that is associated with a specific UID could be tracked down for their whereabouts	No. All communications are implemented wirelessly	6	0	10	8
3	Unauthorized falsification will be ignored by the verification of CRC checksum	No. All communications are implemented wirelessly	0	0	0	0
4	With the UID information in hands, it just extends the possibility to launch all kinds of attack	No. All communications are implemented wirelessly	6	0	0	6
5	Failure of the exchange of challenge information between a registered KEY Fob and the car. The authentication will fail regardless of the users operation	Users will not notice any unusual change until they try to operate the car	8	0	0	8
6	Attacker can get himself/herself authenticated with the recorded challenge information	No. All communications are implemented wirelessly	10	0	8	9
7	Failure of the exchange of response information between a registered KEY Fob and the car. The authentication will fail regardless of the users operation	Users will not notice any unusual change until they try to operate the car	8	0	0	8
8	Without the genuine KEY to correctly encrypt the information, this fake response will be rejected at the car	No. All communications are implemented wirelessly	4	0	0	4

A[1]: Automobile, U[2]: User's Well being, P[3]: Privacy, S[4]: Severity

what is the threshold of the combination? Two groups of guidewords have been combined to consider the deviations in this chapter, however for example, it is not impossible to add back the Secondary Guidewords again into the combination, generating a combination of guidewords from 3 groups to commence such attacks step by step with the primary guidewords, the secondary guidewords, and the tertiary guidewords. Such as attacks in which an attacker Reads the data during the communication, Modifies them into nonsense packets and Floods them back to network; or an attacker firstly Bypasses the authentication mechanism, then fraudulently Authenticates himself/herself as a legal user, and Spoofs the whole network to communicate to or through him/her. Of course, 4 groups, 5 groups or even much more groups are also considerable for the security analysis. In fact, it is proper to say that the more groups combined the more accurate and comprehensive results can we get. We need to consider the combination threshold about where to stop as the future work. Attacks such as session hijacking takes control of the whole system after the authentication is completed. Thus eavesdropping across the whole exchange of challenge and response information is required to succeed by such attack. Therefore, considering the relationship of the two directions such as Read UID and Return UID, also Challenge and Response, it may not be appropriate to deliberately separate them as two completely different objects to conduct the analysis. Integration of such related exchanges of messages should receive fair attention as well, and should be analyzed as a whole. Improvement such as raising the abstraction level of analysis object diagram may be one of the solutions to minimize the current drawback. Thus using not only the sequence diagram as the analysis object, but also introducing other kinds of diagrams to conduct an even comprehensive analysis on all kinds of different levels.

6 Conclusion

This chapter presented the HAZOP-based security analysis technique, with its new guidewords in order to perform exhaustive security analyses during the system designing. We used 8 actions words extracted from the attack taxonomy by CERT as the new guidewords and performed security analysis by applying them to embedded system designs. Although this HAZOP-based technique did succeed in locating some of the security flaws that may eventually cause an incident to the system, leading to the exposure of privacy, or loss of properties. As future works, we plan to discuss problems found in the case study, and develop a tool to support security analysis based on the proposed method.

References

1. Charette, R.N.: This car runs on code. IEEE Spectr. http://spectrum.ieee.org/transportation/ systems/this-car-runs-on-code (2009). Cited 8 Dec 2015
2. IEC 61508 Edition 2.0.: Functional safety of electrical/electronic/programmable electronic safety-related systems, part 1–8. Int. Electrotechnical Comm. (2010). http://www.iec.ch/ functionalsafety/standards/page2.htm. Cited 8 Dec 2015
3. ISO.: 26262 Road vehicles-Functional safety–part 1–9. ISO (2011). http://www.iso.org/iso/ catalogue_detail?csnumber=43464. Cited 8 Dec 2015

4. Koscher, K., Czeskis, A., Roesner, F., Patel, S., Kohno, T., Checkoway, S., McCoy, D., Kantor, B., Anderson, D., Shacham, H., Savage, S: Experimental security analysis of a modern automobile. CAESS (2010). http://www.autosec.org/pubs/cars-oakland2010.pdf. Cited 8 Dec 2015

5. Brooks, R.R., Sander, S., Deng, Juan, Taiber, Joachim: Automobile security concerns, challenges and state of the art of automotive system security. Veh. Technol. Mag. IEEE. **4**(2), 52–64 (2009)

6. Atmel.: Open source immobilizer protocol stack. Atmel (2015). http://www.atmel.com/tools/OPENSOURCEIMMOBILIZERPROTOCOLSTACK.aspx. Cited 8 Dec 2015

7. Tillich, S,, Wjcik, M.: Security analysis of an open car immobilizer protocol stack. Cryptology ePrint Arch. (2012). https://eprint.iacr.org/2012/617.pdf. Cited 8 Dec 2015

8. Pumfrey, D.J.: The principled design of computer system safety analyses. University of York, Department of Compurter Science (1999). https://www.cs.york.ac.uk/ftpdir/reports/2000/YCST/05/YCST-2000-05.pdf. Cited 8 Dec 2015

9. Leveson, G.L.: Safeware: system safety and computers. Addison Wesley Prof. (1995)

10. Dobbing, B., Lautieri, S.: SafSec: integration of safety & security certification, safsec methodology: guidance material. Intell. Syst. ALTRAN Syst. (2006). http://intelligent-systems.altran.com/fileadmin/medias/0.commons/documents/Technology_documents/SafSec_Methodology_Guidance_Material_pdf.pdf. Cited 8 Dec 2015

11. Dobbing, B., Lautieri, S.: SafSec: integration of safety & security certification, safsec methodology: standard. Intell. Syst/ALTRAN Syst. (2006). http://intelligent-systems.altran.com/fileadmin/medias/0.commons/documents/Technology_documents/SafSec_Methodology_Standard_Material_pdf.pdf. Cited 8 Dec 2015

12. Intelligent Systems/ALTRAN Systems.: SafSec: integration of safety & security certification. Intell. Syst/ALTRAN Syst. (2006). http://intelligent-systems.altran.com/en/technologies/security/safsec.html. Cited 8 Dec 2015

13. Young, W.E., Jr.: STPA-SEC for cyber security mission assurance. Eng Syst. Div. Syst. Eng. Res. Lab. (2014). http://psas.scripts.mit.edu/home/wp-content/uploads/2014/03/Young_STAMP_2014_As-delivered.pdf. Cited 8 Dec 2015

14. Raspotnig, C.: Requirements for safe and secure information systems. Department of Information Science and Media Studies, University of Bergen (2014). http://www.uib.no/sites/w3.uib.no/files/attachments/phd_thesis_christian_raspotnig_0.pdf. Cited 8 Dec 2015

15. Howard, J.D., Longstaff, T.A.: A common language for computer security incidents. Sandia Natl. Laboratories (1998). http://cyberunited.com/wp-content/uploads/2013/03/A-Common-Language-for-Computer-Security-Incidents.pdf. Cited 8 Dec 2015

Multimodal Heterogeneous Monitoring of Super-Extended Objects: Modern View

Andrey V. Timofeev and Viktor M. Denisov

Abstract This chapter provides modern view on the super-extended objects monitoring. The monitoring process is being reduced to the detection and classification of targeted events occurred in the vicinity of the controlled object by tracking changes in the internal state of the monitored object and by search for precursors of an environmental change, which can serve as precursors to natural and technological disasters. Suggested approach is based on the multimodal concept of the monitoring object observation, heterogeneous data fusion, detection and classification of targeted events. The approach assumes that different types of physical field are observed simultaneously in real time, data is received from different types of sensors in various rate with different accuracy, with insufficient prior information about distribution probability of targeted signals and background noises. The suggested approach provides stable detection of targeted events, which guarantees upper bounds for probabilities of type I and type II errors. Identification of targeted events type (classification problem) is based on the heterogeneous data fusion methodology. The application results of the proposed approach in the real monitoring system are presented herein.

1 Introduction

The problem of complex monitoring of the super-extended objects has always been of great practical importance. For example, oil and gas pipelines, railways, national frontier are examples of typical super-extended objects, and all of these facilities need to be remotely monitored. Complex monitoring provides solutions for the following tasks:

A.V. Timofeev (✉)
LLP EqualiZoom, Astana, Kazakhstan
e-mail: timodeev.andrey@gmail.com

V.M. Denisov
Flagman Geo Ltd, Saint-Petersburg, Russia
e-mail: denisov.viktor@flagman-geo.ru

© Springer International Publishing Switzerland 2016
E. Pricop and G. Stamatescu (eds.), *Recent Advances in Systems Safety and Security*,
Studies in Systems, Decision and Control 62, DOI 10.1007/978-3-319-32525-5_6

97

- unauthorized activities detection (tie-into a pipeline, excavation in the monitoring object vicinity, pedestrian activity on railways, etc.);
- telemetric status check of technological equipment on the monitoring objects (deformation of pipelines, leaks detection, deformation of building structures);
- control of natural objects that form unified geotechnical-system with monitored object (soil displacement under building foundations, development of karst processes etc.);
- timely detection of technogenic or natural disasters (oil leaks, landslides, train derails, damage of railway tracks), which appear in the monitoring objects vicinity.

Solution for these problems is based on a detection of certain precursors that signal the emergence of targeted events or processes. We will call those precursors a "targeted events". Examples of targeted events: a seismoacoustic vibration accompanying the oil spill from the pipe; a seismoacoustic noise and other symptoms associated with unauthorized attempts of tie-into pipeline or excavation near the railways; seismoacoustic signals associated with pedestrian activity in vicinity of railways; seismoacoustic signals associated with landslides in region of the monitoring object localization.

Existing multimodal solutions for monitoring of super-extended objects use networks of seismic sensors to control the seismoacoustic field near of the object. This approach is becoming expensive if the object's perimeter exceeds 10 km with a system resolution of 10 m. A cost of such system will be 4–5 times more expensive then cost of the monitoring systems using distributed acoustic sensors (DAS) [1, 2]. This is due to a need to provide electrical power and radio communication for each sensor of the network. At the same time the sensors of DAS-monitoring systems have to be installed to meet a special condition: depth not less than 50 cm, offset from the monitoring object up to 5–10 m. These conditions cannot be met in all areas where monitoring objects are situated. Therefore, there is a need for integrated solutions, which means solutions using different types of sensors. Different sensors have divergent sets of specific characteristics, and they measure various physical fields with dissimilar accuracy. It entails the problems with correctness of a heterogeneous information fusion when system detects and classifies targeted events. Additional problems are connected with insufficient prior information about statistical characteristics of the signals and noises.

The approach described in this chapter is intended to provide the multipurpose multimodal monitoring based on concurrent observations of various types of physical fields (seismoacoustic, optical/IR, magnetic etc.). As it was proved by multiple experiments, a joint processing data of these systems significantly improves of the monitoring reliability. Practically effective methods of information fusion are suggested, which designed for guaranteed detection/classification of the targeted events in condition of a priori indeterminacy.

2 Research Objective

Monitoring systems of the super-extended objects are designed to control the operational situation in the vicinity of the monitoring object. The operational situation consists of spatial-time events (STE) flow. The STE's appear in vicinity of monitored object. In common case, STE's can have the dynamic nature; also they can form groups (flows). Obviously, the targeted events' set is a subset of the STE-set. The measurements, which were obtained from sensors of various types, are as the raw data for comprehensive analysis in the monitoring systems. The monitored object represents a distributed system of material assets. The monitoring systems are designed to control the STE's flows in vicinity of these material assets. In monitoring process, we have to detect the STE and classify the STE type (identify targeted events). STE examples are: pedestrian or group of pedestrians in vicinity of railways; technological activities, which are carried out near to material assets of monitored object (operations with heavy equipment near the oil and gas pipelines, operation of maintenance crews on the rail tracks, etc.); train traffic; cars traffic near gas pipelines; unauthorised tie-into pipelines; abrupt soil shift in the area of monitoring object; pedestrians in a vicinity of national frontier. Obviously, almost every STE is a source of seismoacoustic emission source (SES). In this connection, almost every STE may be detected and classified with usage of sensors, which measure the seismic field.

The basic goal of comprehensive monitoring of super-extended objects is the solution of the following tasks: Task "D" (Detection)—detection of the STE; Task "L" (Localization)—estimation of localization of the detected STE with simultaneous estimation of dynamics parameters; Task "C" (Classification)—classification of detected STE by means of assigning it to one of N priori given classes; Task "S" (Object Status Definition: Assessment of Risks)—evaluation of the degree of monitored object operating safety.

Priori N of targeted STE classes (types) are given, and these classes form the set of types (ST). As rule, the power of this set does not exceed a few dozen units. For various application areas ST's can be significantly different, and to have a different features. For example, technological operations on gas pipeline significantly differ to technological operations on railway tracks as in type of seismoacoustic vibrations, as well as in type of visual images. In order to monitor targeted object in area of its localization are installed sensors of different physical nature (DAS, nets of point seismic sensors, nets of inclinometers, far CCTV etc.). Data received from these sensors is heterogeneous and therefore requires a special pre-processing. Each type of sensor generates various types of STE-features, which have different informativeness level and reliability. It is therefore necessary to use a correct fusion procedure of heterogeneous data to provide an effective solution of "D", "L", "C" and "S" tasks. Moreover, solving task "D" is necessary to overcome high degree of uncertainty, which is related to probability characteristics of targeted signals and background noises. This situation is due to the fact that statistical characteristics of background noises are significantly differ around a monitoring object and at the same time, these characteristics have a dynamic nature. All of it represents a significant problem of how the detection subsystem parameters must constantly adjust to changing of background noise characteristics probability. The goal of our research is to suggest effective solutions of tasks "D" and "C", with usage of different types of sensors placed along a super-extended monitoring object. Solution of task "L" may be reduced to an ordinary triangular task. Methods of this task's solution are well known and are not included herein due to the article size restrictions. The task "S" solution is a separate problem, the solution of which is beyond the scope of this study. Suggested approaches for solution of tasks "D", "C" are based on the modern data processing methods, which guarantee high performance and reliability of the monitoring system.

3 Architecture of Heterogeneous Multimodal Monitoring System

In this section we describe general principles of heterogeneous multimodal monitoring system architecture as well as the term "channel" of monitoring system.

3.1 General Principles of Heterogeneous Multimodal Monitoring System Architecture

The method based on high vibrosensitivity of optical fiber is an effective to monitor super-extended objects [2]. In this case, the optical fiber is buried in a vicinity of monitored object, and therefore is influenced to fluctuations of the same seismic field. This fact gives possibility to control a seismoacoustic field near monitored object. So-called C-OTDR monitoring systems [2] belong to this class of systems. Acronym "C-OTDR" is disclosed as "Coherent Optical Time Domain Reflectometer". There are three basic system units: impulse coherent laser, monomode optical fiber (distributed fiber optical sensor: DFOS), and processing unit. An impulse infrared laser sends an infrared coherent stream into a DFOS. Spreading through the DFOS, small part of the injected stream is reflected by the optical fiber imperfections. Reflected part of the stream is called Rayleigh elastic backscattering radiation. Due to a coherency, the reflected stream forms images of chaotic interference at the point of retrieval. Images of chaotic interference are called C-OTDR-speckles structures or simply speckles. These speckles are highly sensitive to vibrations and reflect even weak seismic vibrations which appear near of the DFOS. In simple words, C-OTDR-speckles "vibrates" by reason of initial vibration of seismoacoustic emission source which creates seismic waves in vicinity of DFOS (and near monitored object at the same time). So, these waves are initial reason of C-OTDR-speckles vibration, because of changes in local refractive index of the optical fiber. STE's leaves traces in seismoacoustic field, and STE's are seismoacoustic emission sources (SES) in this case. Seismoacoustic vibrations from those SES's are propagated in the ground, inducing secondary vibrations of C-OTDR-speckles with the same frequency. It is important to mention that Time-Frequency Characteristics (TFC) of C-OTDR-speckle vibrations are similar to TFC of initial SES. Thus seismoacoustic pressure from SES causes a refractive index changing. This in turn causes the C-OTDR- speckles dramatically change. This change forms a useful signal eventually. This change forms a useful signal eventually. After that, during Feature Extraction Stage, we retrieve the desired features of the useful signal. These features are used for solution of detection and classification tasks. A simplified scheme of the C-OTDR-system is shown on Fig. 1.

In addition to the C-OTDR-sensor, class DAS includes several types of sensors, which use different optical effects. C-OTDR-sensor uses Rayleigh backscatter. The Brillouin fiber sensors use Brillouin Scattering. Both of these sensors belong to a DAS class. In practice, there are always sites of the object perimeter, where it is impossible to use the optical fiber or where the fiber-optic based technologies are non-effective. For example, part of the oil pipeline may be placed on soil with high acoustic impedance (silt, sand) therefore seismoacoustic waves of STE's will not reach to optical fiber. Thus we will not have information about these STE's. For detection and classification of these STE's we must use sensors of another physical type (modality), for example, far CCTV or sensors to monitor gamma-ray background. Another case is connected with situation, when we need to control a development of karst processes near monitored objects. Here, methods with optical fiber usage are ineffective, and it is better to use net of inclinometers. Thus we have objective situation when the monitoring system must to be multimodal, at the same

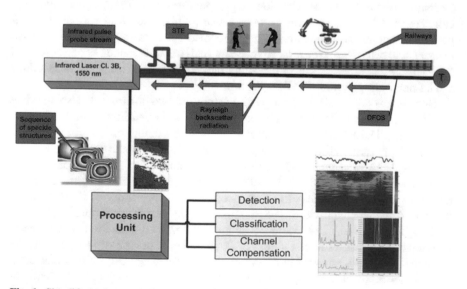

Fig. 1 Simplified scheme of C-OTDR monitoring system

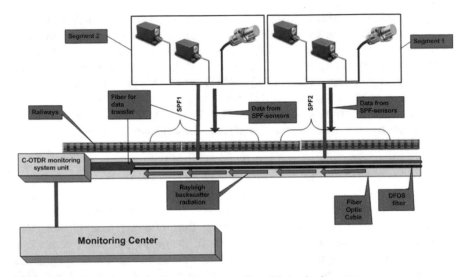

Fig. 2 Simplified scheme of a heterogeneous multimodal monitoring system

time architecture of monitoring system must be heterogeneous to answer real monitoring challenges. On the design stage of monitoring system, we must distinguish between object perimeter fragments that need special supervision or certain key features. For example, when designing a monitoring system intended to monitor railroad tracks, we should focus on fragments of object perimeter that are close to the places of probable displacement of rocky ground. We will denote such fragments by a SPF. The architecture of heterogeneous multimodal monitoring system consists of subnets of various types point-sensors. These subnets we will call "segments". Each segment includes set of point-sensor with various physical natures' (geophones, gamma-ray sensors, camcorders, inclinometers etc.). These sensors are placed around respective SPFs and we will name these sensors as SPF-sensors. The Fig. 2 shows the simplified scheme of a heterogeneous multimodal monitoring system (HMM-system).

Here fiber-optic cable is used as DAS (one or two fibers) and also for data transfer (rest fibers). Of course, this cable should be buried in a vicinity of monitored object. The HMM-systems are multichannel by definition and for system of such kind the term "channel" is very important. This common term implies two meanings: DAS-channel and Information channel for getting data from SPF-sensors. The DAS-channel is virtual one. De facto it is a small part of DFOS, data from which corresponds to seismoacoustic field in vicinity of it. The size of this part (DAS-channel width) is determined by the length of the probe pulse. In practice the DAS-channel width is about 2–10 m and namely this parameter determines a spatial resolution ability of DAS-systems. In case of SPF-sensors, the "channel" is simple physical information channel (fiber-optical, twisted pair etc.), and this term has clearly physical meaning. On level of data processing, each of these channel types viewed as tool for receiving data from particular point of the object. At the same time, each of those tools has individual indicators of quality, the rate of measurement, and accuracy. All of those indicators we take into account for solution of tasks "D", "L", "C" and "S".

3.2 Information Channels of Heterogeneous Multimodal Monitoring System

Indexes of channels HMM-system in conjunction form a set $Z = \{1, 2, \ldots\}$. In every channel $j \in \mathbf{Z}$, $S_j(t)$ is a measurement of the seismoacoustic field as function of time. We will consider two general types of sensors: point-sensors and DAS. In case of point-sensor, each channel $j \in \mathbf{Z}$ is characterized by its coordinates (it is coordinates of point-sensor). In case of DAS, every channel represents the sector of DFOS of length N meters. Coordinates of every DAS-channel $j \in \mathbf{Z}$ are fully defined by the coordinates of its beginning $b(j)$ and ending $e(t)$ points. The 2-tuple $(b(j), e(j))$ is called boundaries (start and end) of the channel j. If two channels match at least one of the boundaries, these channels are called adjacent. A group of adjacent channels (GAC) is the set of channels, each of which is adjacent to at least one of the GAC. The boundaries of all channels are given and form the tuple $\mathbf{X}(\mathbf{Z}) = \langle x_1, x_2, \ldots \rangle$. For jth channel we have $(b(j), e(j)) = (\langle \mathbf{X}(\mathbf{Z}) \rangle_j, \langle \mathbf{X}(\mathbf{Z}) \rangle_{j+1})$, here $\langle \mathbf{X} \rangle_j$ is jth component of the tuple $\mathbf{X}(\mathbf{Z})$. STE's leaves traces in seismoacoustic field, and STE's are the cause of SES appearance in this case. Next, seismoacoustic wave propagating from SES, reaches a certain GAC with delays proportional to a distance from the location of the SES to a particular channel of GAC. The specific composition of GAC determined by the location SES, its energetic power, distance of up to FOS, as well as the parameters of the wave propagation medium. We call the GAC, which turned out under the influence by the seismoacoustic waves from SES, the detecting GAC (DGAC). Observations are made at successive times, which form a set $T = \{t_1, t_2, \ldots\}$, $\forall i > 0 : t_{i+1} - t_i = \Delta t > 0$. Thus, the observations are form the following sets $\mathbf{S}_j = \{S_j(t) | t \in T\}$, $j \in DGAC$. All channel and inter-channel statistics are defined on the intervals of duration Δ. Each of those intervals contains z of the discrete observations.

4 The Guaranteed Detection of the Spatial-Time Events in Multi-channel Monitoring Systems

In this section we describe two approaches to guaranteed detection of STE in multi-channel monitoring systems. First of them is based on simultaneously data processing in several channels. This approach can be useful for DAS-sensors, as well as for point-sensors. Second approach is based on separately data processing in every channel. This method is useful for both types of sensor too, but this method requires less computational resources.

4.1 The Robust Guaranteed Detection of the Spatial-Time Events with Usage of Simultaneously Data Processing in Several Channels

For convenience, we will describe suggested method on example of DAS monitoring system (DAS-MS). In DAS-MS, STE's regarded as SES's. Specificity of the modern DAS monitoring systems is such that any two channels are statistically independent only if no SES, the elastic vibration from which affects the speckle patterns of these channels simultaneously. If such the SES exists, those channels are become statistically dependent. In this case we can speak about a group of SES, seismoacoustic waves from which will affect the FOS channels simultaneously and compositionally. Obviously, the observations of any two channels of the DGAC are statistically dependent. We call the area the sensitivity of the monitoring system the area Ω which situated in vicinity of FOS and with the appearance inside which one or a group of SES, the observations $S_j, j \in DGAC$, of the appropriate DGAC will be abruptly change their statistical characteristics. In other words, observations $S_j, j \in DGAC$ will are mutually dependent after appearance inside Ω one or a group of SES.

4.1.1 Requirements for the Decision Procedure

Let us denote

- τ is the moment of abrupt change of the observations distributions, which happened because of appearance the SES in Ω; actually, τ is a point of appearance the signals from the SES in DGAC or **change-point moment**;
- hypothesis H_0: in the region Ω do not SES (background model);
- hypothesis H_1: in the region Ω is at least one SES (signaling model);
- $\alpha \in]0, 1[$ is a predetermined upper bound for the probability of making type I errors;
- $\beta \in]0, 1[$ is a predetermined upper bound for the probability of making type II errors;
- $\Delta(t) = [t - z, t]$ is the interval for calculation of the speckle-structures statistical characteristics (speckle-metrics), where z is the number of the discrete observations inside of the interval $\Delta(t)$; during monitoring the interval $\Delta(t)$ is shifted by an amount Δt along the time axis T;

- $$\widetilde{S}_j(t, t+z) = \{S_j(t), S_j(t+1), \ldots, S_j(t+z)\} \subseteq \mathbf{S}_j;$$

- $$\overline{\widetilde{S}}_j(t_1, t_2) = \sum_{t=t_1}^{t_2} S_j(t)(t_2 - t_1)^{-1}; \overline{\overline{S}}_j(t_1, t_2) = \sum_{t=t_1}^{t_2} \left(S_j(t) - \overline{\widetilde{S}}_j(t_1, t_2)\right)^2;$$

- $$\overline{\overline{S}}_{ji}(t_1, z, \delta) = \overline{\overline{S}}_j(t_1, t_1 + z)\overline{\overline{S}}_i(t_1 + \delta, t_1 + z + \delta);$$

- $$U(z|t, t_1, j) = \left(\widetilde{S}_j(t) - \overline{\widetilde{S}}_j(t_1, t_1 + z)\right);$$

- $$r^{(i,j)}(t_1, t_1 + z|\delta) = \sum_{t=t_1}^{t_1+z} U(z|t, t_1, j)U(z|t+\delta, t_1+\delta, j)\left(\overline{\overline{S}}_{ji}(t_1, z, \delta)\right)^{-0.5};$$

- $\Sigma(\mathbf{Z}) = \bigcup_{j \in DGAC} \mathbf{S}_j; \rho(t|\Sigma(\mathbf{Z}), \mathbf{\Delta}(t)) \subseteq R^1$ is some stochastic function, which we will call the **signaling function**; this function is defined on the interval $\mathbf{\Delta}(t)$ and depends on the $\Sigma(\mathbf{Z})$ so that: $\mathbf{E}(\rho(t|\cdot)|H_0) = 0, \mathbf{E}(\rho(t|\cdot)|H_1) > 0, \mathbf{E}(\rho^2(t|\cdot)|H_0) < \infty.$

Watching $\mathbf{S}_j = \{S_j(t)|t \in T\}, j \in \mathbf{Z}$, need to define a decision function $W(t, z, \rho, \alpha, \beta) \in \{0, 1\}$, that depends on the $\rho(\cdot), \mathbf{\Delta}(t) = [t - z, t]$, and on the parameters α, β such that

- $$P[W(t, z, \rho, \alpha, \beta) = 1|\tau \notin [t - z, t]] \leq \alpha,$$

- $$P[W(t, z, \rho, \alpha, \beta) = 0|\tau \in [t - z, t]] \leq \beta.$$

Thus, it has been tasked interval estimation of the change-point τ (τ is the moment of the appearance the SES in Ω). The solution should guarantee the pre-determined upper bounds for the probabilities of making type I (α) and type II (β) errors. In this formulation, the problem of detection SES reduces to problem of interval estimation of **change-point** τ.

4.1.2   Selecting a Robust Signaling Function ρ

By definition the signaling function $\rho(t|\Sigma(\mathbf{Z}), \Delta(t))$ abruptly changes its average value (towards increase) at the moment τ. It's desirable to the probability distribution of the ρ had been robust to outliers in the observations $\Sigma(\mathbf{Z})$. By the condition of the problem statement, vectors, $\widetilde{S}_i(t+\delta, t+z+\delta)$ satisfy the conditions of Theorem 8.1 [4], p. 204, if $t+z+\delta < \tau$. Following this theorem, $\forall_{i,j,\delta}\{\mathbf{E}(r^{(i,j)}(t, t+z|\delta)) = 0, \mathbf{E}(r^{(i,j)}(t, t+z|\delta))^2 = (z-1)^{-1}\}$. As the problem statement dictates we can use the following function as a signaling one:

$$\rho(t|\Sigma(\mathbf{Z}), \Delta(t)) = \sup_{i,j \in DGAC}\left[\sup_{\delta}\left(r^{(i,j)}(t, t+z|\delta)\right)\right]. \qquad (4.1)$$

However, this function is not robust to the statistical anomalies of observations. The robust estimates of the correlation coefficient were considered in [3–7]. In Sect. 8.3 [5], a powerful approach was suggested to obtain the robust estimates. Following [5], when calculating $r^{(i,j)}(t, t+z|\delta)$, instead $\widetilde{S}_j(\cdot), \widetilde{S}_i(\cdot)$ uses some $\mathbf{u}\left(\widetilde{S}_j(\cdot)\right), \mathbf{v}\left(\widetilde{S}_i(\cdot)\right)$, which were calculated from $\widetilde{S}_j(\cdot), \widetilde{S}_i(\cdot)$ respectively, according to next five rules:

1. $\mathbf{u} = \Psi\left(\widetilde{S}_j(\cdot)\right), \mathbf{v} = \Xi\left(\widetilde{S}_i(\cdot)\right)$;
2. Ψ, Ξ commute with permutations of the components of $\widetilde{S}_j(\cdot), \mathbf{u}$ and of $\widetilde{S}_i(\cdot), \mathbf{v}$;
3. Ψ, Ξ preserve a monotone ordering of the components of $\widetilde{S}_j(\cdot), \widetilde{S}_i(\cdot)$;
4. $\Psi = \Xi$;
5. $\forall a > 0, \forall b, \exists a_1 > 0, \forall b_1, \forall \mathbf{x}\Psi(a\mathbf{x}+b) = a_1\Psi(\mathbf{x})+b_1$.

In the following two examples (from Sect. 8.3 [5]) all five requirements hold:

- the classical Spearman rank correlation between $\widetilde{S}_j(\cdot)$ and $\widetilde{S}_i(\cdot)$;
- the quadrant correlation between $\widetilde{S}_j(\cdot)$ and $\widetilde{S}_i(\cdot)$.

If the $r^{(i,j)}(t, t+z|\delta)$ was calculated by means of the classical Spearman rank, we will denote this as $r_{sr}^{(i,j)}(t, t+z|\delta)$. And if the $r^{(i,j)}(t, t+z|\delta)$ was calculated using the quadrant correlation, we will use the following notation $r_{qc}^{(i,j)}(t, t+z|\delta)$. Both of them $r_{sr}^{(i,j)}(t, t+z|\delta)$ and $r_{qc}^{(i,j)}(t, t+z|\delta)$ are robust to the statistical anomalies of observations. Therefore, the corresponding signaling functions

$$\rho_{sr}(t|\Sigma(\mathbf{Z}), \Delta(t)) = \sup_{i,j \in DGAC}\left[\sup_{\delta}\left(r_{sr}^{(i,j)}(t, t+z|\delta)\right)\right]$$

and

$$\rho_{qc}(t|\Sigma(\mathbf{Z}),\Delta(t)) = \sup_{i,j\in DGAC}\left[\sup_{\delta}\left(r_{qc}^{(i,j)}(t,t+z|\delta)\right)\right]$$

are robust too. It easy to see:

$$\forall t<\tau : \left[\mathbf{E}\rho_{sr}(t|\cdot) = 0, \mathbf{E}\rho_{sr}^2(t|\cdot) = 1/(z-1)\right] \qquad (4.2)$$

$$\forall t<\tau : \left[\mathbf{E}\rho_{qc}(t|\cdot) = 0, \mathbf{E}\rho_{qc}^2(t|\cdot) = 1/(z-1)\right].$$

Thus, functions ρ_{sr} and ρ_{qc} may be used as the robust signaling functions (RSF). The RSF we will denote as s $\rho_{rb} \in \{\rho_{sr}, \rho_{qc}\}$.

4.1.3 Guaranteed Detection Method of the SES

As it follows from (4.2), until the moment τ the expectation of the RSF is zero. Under influence of elastic vibrations from SAE, expectation of the RSF abruptly changes towards increase, because the observations at once several channels of DGAC become statistically dependent. In other words, at the moment τ the model of RSF will get change-point of its probabilistic properties compared with the background model H_0. Let us describe the method for interval estimation of the moment τ.

Remark 1 If $W(t,z,\rho,\alpha,\beta) = 1$ the $[t-z,t]$ will be confidence interval for moment τ. In the case of realization the event $W(t,z,\rho,\alpha,\beta) = 1$ decision will be taken that $\tau \in [t-z,t]$. Essentially—it is a fact of detection SAE. And it is guaranteed the predetermined upper bounds for the probabilities of making type I (α) and type II (β) errors.

Properties of the proposed method are described in the following theorem:

Theorem 4.1 *Let*

1. $\exists C > 1 : \mathbf{Var}(\rho_{rb}(t|\Sigma(\mathbf{Z}),\Delta(t))) \leq Cz^{-1}$;
2. $\exists \theta > 0 : \inf_{t\geq\tau} \mathbf{E}\rho_{rb}(t|\Sigma(\mathbf{Z}),\Delta(t)) \geq \theta$;;

3. $W(t,z,\rho,\alpha,\beta) = \begin{cases} 1, & \text{if } \rho_{rb}(t|\Sigma(\mathbf{Z}),\Delta(t))/\theta \geq b \\ 0, & \text{if } \rho_{rb}(t|\Sigma(\mathbf{Z}),\Delta(t))/\theta < b. \end{cases}$

Here $b = \left(\frac{\beta}{\alpha C^2}\right)^{0.5}\left(1-\left(\frac{\beta}{\alpha C^2}\right)^{0.5}\right)^{-1}$.

Then, if $z = 1 + \frac{C^2}{\theta^2\alpha} - \frac{2C}{\theta^2\sqrt{\alpha\beta}} + \frac{1}{\theta^2\alpha}$, *the following statements are true:*

$$P[W(t, z, \rho, \alpha, \beta) = 1 | \tau \notin [t - z, t]] \leq \alpha$$
$$P[W(t, z, \rho, \alpha, \beta) = 0 | \tau \in [t - z, t]] \leq \beta.$$

Remark 2 The constant θ has following sense: the algorithm will detect those SAE, for which the increment of the value $\mathbf{E}\rho_{rb}(t|\boldsymbol{\Sigma}(\mathbf{Z}), \boldsymbol{\Delta}(t))$ when $t - z \geq \tau$ will exceed the value θ. Thus, parameter θ defines the sensitivities of the detection algorithm.

Proof of Theorem 4.1 Consider the representations

$$\rho_{rb}(t|\cdot))/\theta = \begin{cases} m(t), t < \tau \\ \mathbf{E}\rho_{rb}(t|\cdot)/\theta + m'(t), t - z \geq \tau \end{cases}$$

In view of (4.2) we have:

$$\mathbf{E}m(t) = 0, \mathbf{E}m^2(t) = 1/(\theta^2(z - 1)). \tag{4.3}$$

Using Chebyshev's inequality, we write:

$$P[W(t, z, \rho, \alpha, \beta) = 1 | \tau \notin [t - z, t]] =$$
$$P[\rho_{rb}(t|\cdot)/\theta > b | \tau \notin [t - z, t]] \leq \tag{4.4}$$
$$P[|m(t)| > b] \leq (\theta^2(z - 1)b^2)^{-1}.$$

When $\tau > t - z$ from condition 2 we have $\mathbf{E}\rho_{rb}(t|\boldsymbol{\Sigma}(\mathbf{Z}), \boldsymbol{\Delta}(t))/\theta \geq 1$. Therefore, taking into account the first condition of the theorem, we can write:

$$P[W(t, z, \rho, \alpha, \beta) = 0 | \tau \in [t - z, t]] =$$
$$P[\mathbf{E}\rho_{rb}(t|\cdot)/\theta + m'(t) < b | \tau \in [t - z, t]] \leq$$
$$P[|\mathbf{E}\rho_{rb}(t|\cdot)/\theta| - |m'(t)| < b | \tau \in [t - z, t]] \leq$$
$$P[|m'(t)| > |\mathbf{E}\rho_{rb}(t|\cdot)/\theta| - b] \leq \tag{4.5}$$
$$P[|m'(t)| > 1 - b] \leq C^2/(z\theta^2(1 - b)^2) \leq$$
$$C^2/((z - 1)\theta^2(1 - b)^2).$$

Substituting in (4.4) and (4.5) the values of z and b, as defined in the condition of the theorem immediately confirm the truths of the allegations are proved. The theorem is proved.   ∎

The approach described in this report is used for the detection of SES in real C-OTDR monitoring system (C-OTDR-MS). Parameters of the C-OTDR-MS: the probe pulse duration: 50–150 ns; frequency sensing: 2–8 kHz; the probe signal power—15 mW; laser wavelength: 1550 nm. Table 1 contains results of SES's detection. Here «Distance» is an average distance at which the given class of SES was detected, P_I—is detection error of the type I; P_{II}—error of the type II. Parameters of the detection system were such: z = 25, $\theta = 0.2$, $\alpha = 0.1$, $\beta = 0.1$. The parameter C was estimated experimentally: C $\sim$ 1.3. Data are presented for

Table 1 The practical detection results

Type of SES	Distance (m)	P_I	P_{II}
"hand digging the soil"	10	0.09	0.08
"chiselling ground scrap"	5	0.1	0.1
"walking man"	10	0.07	0.09
"running man"	15	0.09	0.08
"passenger car"	25	0.06	0.1
"truck"	35	0.07	0.09
"heavy equipment excavator"	50	0.02	0.01
"easy excavation equipment"	40	0.03	0.02

rocks cemented soil. The results in Table 1 show sufficiently high practical effectiveness of the described SES detection system. We have to note this method is very demanding to computational resources; therefore it may only be used in systems with a small number of channels.

4.2 Method of Adaptive Sequential Real-Time Detection of Spatial-Time Events in Multichannel Monitoring Systems

In case of super-extended object monitoring, the conditions of observation are dramatically different at different times and different places. These circumstances strongly influence sensors systems. The influence implies an increase in Type I and Type II errors. While noise level may dramatically be different in various time intervals for one and the same channel, the industrial noise power and its spectral characteristics as a rule are stable for extended periods of time (no less than a few minutes or sometimes even hours). Unlike the industrial noises, targeted signals have high power and short duration (no more than a few minutes). So, targeted signals have a shorter stability period with respect to stability period of noises. Using this assumption, we can build an adaptive real-time detector which will guarantee prescribed level for Type I and Type II errors. This approach will be described in this section on an example of C-OTDR-MS with a high number of channels (more than 20,000). Suggested method contains two phases: (a) phase adaptation to background noises, and (b) phase of SES's detection. Both phases are based on a sequential analysis approach. Let us denote: τ_j is random moment time. So, τ_j is the moment of abrupt change of the observations distributions in jth channel; this change happened due signal appearance; actually, τ_j is the change-point moment [8, 9] of observation distributions; t_0 is time of observation start; h is the sample size for initial adaptation to noise; P_c—confidence coefficient, $0 < P_c < 1$; $\Delta(h) = 2((1 - P_c)h)^{-0.5}$; hypothesis H_0: in channel do not signal (background model); hypothesis H_1: in channel is signal (signaling model); δ—desired size of confidence interval for background model; $\alpha \in (0,1)$ is a predetermined upper bound for the probability of making type I errors; $\beta \in (0,1)$—is a predetermined upper bound for the probability of making type II errors.

4.2.1 SES's Detection Problem Set and Requirements for the Decision Procedure

For each channel $j \in \mathbf{Z}$ observations are described by following expressions:

$$\forall t < \tau_j, t \in \Delta : S_j(t) = \theta_j + \sigma_j(t)\varsigma_j(t)$$

$$\forall t \geq \tau_j, t \in \Delta : S_j(t) = \theta_j + \sigma_j(t)\varsigma_j(t) + \theta_{s,j}(t) + \pi_j(t)\xi_j(t).$$

Here

- $\{\varsigma_j(t)\}, \{\xi_{s,j}(t)\}$ are mutually independent random variables, $\mathbf{E}\varsigma_j(t) = 0, \mathbf{E}\varsigma_j^2(t) = 1, \mathbf{E}\xi_{s,j}(t) = 0, \mathbf{E}\xi_j^2(t) = 1, \sigma_j \leq L_N, \pi_j \leq L_S$; the constants L_N, L_s are given; noise parameters $\{\theta_j\}$ are priori unknown.
- $\forall_{i \neq j} \theta_i \neq \theta_j, \Xi_{s,j}(t) = \theta_{s,j}(t) + \pi_j(t)\xi_j(t)$ is equation of target signal in jth channel, $\theta_{s,j}(t) > 0$.

The research objective is to build a **SES's detection procedure**, which will guarantee the prescribed level for Type I (α) and Type II (β) errors. In solving this problem for each channel we can use observations of this channel only. So, we do not have a possibility to use cross-channel information, in contrast to method of 4.1.3. This is due presence of huge number of channels (more than 20,000), data from which we have to process in real time.

4.2.2 Method of Adaptive Sequential Real-Time SES's Detection

During adaptation phase of background model parameters are estimated. In frame of suggested conception we interest in parameters $\{\theta_j\}$ estimation. We need to know the one-sided confidence upper bounds for parameters $\{\theta_j\}$ with given confidence coefficient P_c. For solution of this task we will use the sequential analysis. Let us consider the following simple statistic:

$$\overline{\overline{\theta}}_j(t_0, h) = \sum_{p=t_0}^{t_0+h} S_j(p)/(L_N h) + ((1 - P_c)h)^{-0.5}.$$

The $\overline{\overline{\theta}}_j(t, h)$ is non-parametric confidence upper bound for θ_j, and it is easy to see that $\mathbf{P}(\theta_j \leq \overline{\overline{\theta}}_j(t_0, h)) \geq P_c$ and $\forall_j P\left(\lim_{h \to \infty} \overline{\overline{\theta}}_j(t_0, h) \to \theta_j\right) = 1$. So, interval

$(t_0, t_0 + h)$ is called initial interval adaptation to noise (IIAN); calculation of $\overline{\overline{\theta}}_j(t, h)$ we will call as "adaptation to noise" (AN-procedure). Let us consider following cycle statistics:

$$Y_j(t|h, H) = (H(\Delta(h) + \varepsilon))^{-1} \left(\sum_{k=t-z}^{t-1} \frac{S_j(k)}{L_N^2 + L_S^2} \right) - \frac{\overline{\overline{\theta}}_j(t_0, h)}{\Delta(h) + \varepsilon} + \frac{\alpha' S_j(t)}{(\Delta(h) + \varepsilon)H}.$$

$$z = \inf\left\{ t \geq t_0 + h | t \geq H(L_N^2 + L_S^2) \right\}, \alpha' = H - (z - 1)/(L_N^2 + L_S^2), \varepsilon > 0$$

Statistics $Y_j(t|h, H)$ are defined on the sequence of the intervals:

$$U(z, t_0) = \{u(t_i, z)|i \geq 1, u(t_i, z) = (t_i, t_i + z + 1), t_i = t_{i-1} + z + 1, t_i \geq t_0\}.$$

Those cycle statistics $Y_j(t|h, H)$ will be used for guarantee detection of signals by reducing the task of detection signals to the task of the moment τ interval estimation. As confidence interval we will consider any interval $u(t_i, z)$ from sequence $U(z, t_0)$. Once $Y_j(t_i + z + 1|h, H)$ is calculated, we have to decide what is right: $\tau \in u(t_i, z)$ or $\tau \notin u(t_i, z)$. Simply put, the accuracy of moment τ estimation is z. The conditions, which guarantee a prescribed reliability of the algorithm, are determined by the following

Theorem 4.2 *Let*

1. $1 - P_c < \alpha, 1 - P_c < \beta$.
2. $\forall_j \exists \varepsilon > 0 : \underset{t \geq \tau}{Inf}\ \theta_{s,j}(t) \geq 2\Delta(h) + \varepsilon$.

3. $H = P_c \left[(\beta - 1 + P_c)(1 - b)^2 (\Delta(h) + \varepsilon)^2 \right]^{-1}$, *where* $b = \left(\frac{cm+1}{c+1} \right), c = \left(\frac{\alpha-1+P_c}{\beta-1+P_c} \right)^{0.5}$,
   $\mathbf{m} = \frac{\Delta(h)}{\Delta(h) + \varepsilon}$.

In this case, if the decision rule will be defined by the following way

$$R(b) = \begin{cases} if\ Y_j(t|h, H) \geq b \text{ then } H_1 \text{ is true in jth channel} \\ if\ Y_j(t|h, H) < b \text{ then } H_0 \text{ is true in jth channel} \end{cases},$$

then next inequalities will be true for prescribed α u β:

1. $\mathbf{P}\big(Y_j(t_i + z + 1|h, H) < b | H_1 : \tau \in u(t_i, z)\big) \leq \beta$
2. $\mathbf{P}\big(Y_j(t_i + z + 1|h, H) \geq b | H_0 : \tau \notin u(t_i, z)\big) \leq \alpha$.

The proving of this theorem is similar to proving of Theorem 4.1. The approach described in this section is used for the detection of SES's in a real C-OTDR-MS, which has been described in 4.1.3. Table 1 contains the results of detection of SES's. Here «Distance» is an average distance at which the given class of SES was detected, P_I—is a detection error of type I; P_{II}—an error of type II; "Type of Noise" is the type of the background industrial noise (there are two types: "noise of water

Table 2 The practical detection results

Type of SES	Distance (m)	P_I	P_{II}	Type of noise
"hand digging the soil"	10	0.15	0.09	Type 1
"chiseling ground scrap"	5	0.18	0.09	Type 2
"walking man"	10	0.19	0.09	Type 2
"cutting frozen soil"	15	0.14	0.1	Type 1
"train"	20	0.0	0.0	Type 2

drain" (type 1), and "noise from the diesel generator" (type 2). Parameters of the detection system were as follows: $h = 200$ (since update rate of models is 20 Hz, the adaptation interval length is 10 s), $\alpha = 0.2, \beta = 0.1$. Table 2 shows sufficiently high practical effectiveness of the described SES detection system.

5 Classifications of Spatial-Time Events

In practice, STE leaves traces in different types of physical fields. Therefore, for STE classification we can use sensors of different types, each of them correspond to respective physical field. Every sensor type implies usage of respective features, which describe the STE, and which can be used for classification. We will not discuss methods of feature selection; we assume that for each sensor type the respective feature set has already been formed. For example, CCTV-system design took into account fact that shape and color features are more efficient in contrast with dynamics features [10]. The shape/color characteristics: Scale Invariant Feature Transform (SIFT) [11], Color SIFT [12], Histogram of Oriented Gradient (HOG) [13], and other. The dynamics features: Space-Time Interest Points (STIP) [14], Dense Trajectories [15]. In case of C-OTDR monitoring system can use the tandem LFCC (Linear-Frequency Spaced Filterbank Cepstrum Coefficients)-GMM (Gaussian mixture model) is the most effective feature for the SES classification [16]. LFCC-GMM-vectors with dimension 1024 were used as C-OTDR features. Here LFCC's are defined for speckle-structures of particular C-OTDR channels.

In this section we consider methods of information fusion, which allow effectively taking into account the heterogeneous data of different sources in process of STE classification.

5.1 The Multimodal Algorithm of STE Classification

In order to provide functionality of the multimodal monitoring system different subsystems were trained together, as group of independent classifiers on the same labeled data. For example, for case of bimodal monitoring system [16] the CCTV and C-OTDR subsystems were trained together with good result. Let us denote φ_j—jth feature; K—number of features; m—number of STE classes; indexes of those classes form a set $\mathbf{I}$; $\mathbf{x}(t)$ is a generalized super-vector of a channel measurements; super-vector $\mathbf{x}(t)$ combines data of all sensors at the time moment t, $\mathbf{x}(t) = (x_1(t), x_2(t), \ldots, x_K(t))$, $\forall x_j(t) \in R^{d_j}$, $\mathbf{x}(t) \in R^D$, $D = \prod_{j=1}^{K} d_j$, $x_j(t)$ is data of jth sensor at the time moment t, R^{d_j} is d_j-dimensional feature space of jth feature. It's obvious that each feature φ_j is a function of argument $x_j(t)$; so we can write $\varphi_j(x_j(t))$. On training stage, each class represents with N samples. $\{(\varphi_j(x_j(t_l)), y_l) | l = 1, \ldots, N\}$—training sample set for jth feature, t_l—time of getting l-sample. Here and further, $y_l \in \mathbf{I}$. Following the conclusions of [17] as algorithm of STE classification was used by a so called multiclass v-LPBoost [18], built as a linear convex hull of Lipschitz classifiers. This method steadily works even at a small training sample size [17]. As the Lipschitz classifiers have used conventional SVM (Support Vector Machine) [19]. We denote $f_{j,i}(\cdot | \alpha_{j,i}, b_{j,i}) \equiv f_{j,i}(\cdot)$—SVM classifier, which corresponds to φ_j feature (jth feature space), and to ith STE class ($i \in \mathbf{I}$); $(\alpha_{j,i}, b_{j,i})$—parameters of jth SVM classifier, those parameters are subject to setting up on training stage. So, we have the set of classifiers $\{f_{j,i}(\cdot | \alpha_{j,i}, b_{j,i}) | i \in \mathbf{I}, j = 1, \ldots K\}$.

To solve the multiclass classification SES problem those SVM-classifiers were trained by well-known scheme **one-against-all** (for according feature spaces). According to the concept **one-against-all** every class i is separated from the other classes by use the corresponding classifier $f_{j,i}(\cdot | \alpha_{j,i}, b_{j,i})$ in the respective feature space. All these SVM-classifiers $f_{j,i}(\cdot | \alpha_{j,i}, b_{j,i})$ are built based on the product Bhattacharya kernels [20]. Optimization of the classifiers parameters $(\alpha_{j,i}, b_{j,i})$ was made by use of usual cross-validation (CV) scheme.

A multimodal discriminant function of v-LPBoost-classifier [17], has the following simple form:

$$F(\mathbf{x}(t)) = Arg \max_{i \in \mathbf{I}} \left(\sum_{j=1}^{K} \beta_j f_{j,i}(\varphi_j(x_j(t_l)) | \alpha_{j,i}, b_{j,i}) \right).$$

The training phase comes down to an optimal choice of parameters $\{\beta_j\}$. This choice is performed by using standard optimization method (**linear programming**) according to the following scheme:

$$\min_{\beta, \xi, \rho} \left(-\rho + \tfrac{1}{vN} \left(\sum_{k=1}^{N} \xi_k \right) \right), \text{ under the condition:}$$

$$y_k \sum_{j=1}^{K} \beta_j f_{j,y_k}(\varphi_j(x_j(t_l)) | \cdot) - \arg \max_{y_p \neq y_k} \left(\sum_{j=1}^{K} \beta_j f_{j,y_p}(\varphi_j(x_j(t_l)) | \cdot) \right)$$

$$+ \cdots + \xi_k \geq \rho, k = 1, \ldots N, \sum_{j=1}^{K} \beta_j = 1, \beta_j \geq 0, j = 1, \ldots K.$$

Here ξ—slack variables, v—regularization constant, which is chosen using CV. The training was carried out with the advice of [17]. Thus, the two stage scheme was used to avoid the biased estimates. In [17] this scheme is described in detail.

5.2 The STE Classification Algorithm Application Results

The multimodal algorithm of STE classification showed good performance and stability when used in a bimodal surveillance system [16]. This system is using measurements of two types of physical fields: seismoacoustic and optical/infrared. In this case both the C-OTDR-MS and the Far CCTV (FCCTV) systems are used together. This bimodal surveillance system is installed at the special test area of Kazakhstan Railroads (near Astana). The C-OTDR-MS distributed fiber optic sensor (DFOS) length is 1500 m. This DFOS is buried near the railroad in depth of 50–100 cm. The quality of the FCCTV video stream is 25 fps at a resolution of 704×576 pixels. The system works in real time, and its goal is to monitor an operational situation in vicinity of railroad ballast prism. The targeted STE set is: "pedestrian", "technological activities near the ballast prism", "passenger-train", "cargo-train", "shunting-train", and "animals". As it was mentioned above, in

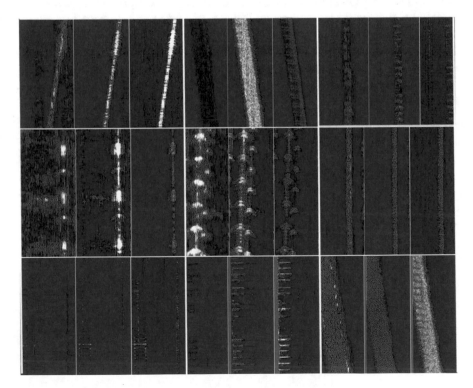

Fig. 3 Waterfall-charts of different SES types

Table 3 Performance Characteristics of the Bimodal Surveillance System

Type of SES	Distance from FCCTV (m)	α	β
Technological activities	200	0.02	0.03
	500	0.03	0.04
Pedestrian	200	0.07	0.08
	500	0.1	0.09
Shunting-train	200	0.06	0.03
	500	0.05	0.04
Cargo-train	200	0.01	0.01
	500	0.01	0.01
Passenger-train	200	0.01	0.01
	500	0.02	0.01

C-OTDR-MS the spatial-time events regarded as SES's. For example, the Fig. 3 represents so-called waterfall-charts of typical seismoacoustic emission sources.

Here the axis Y relates to time, the axis X relates to distance, color encodes intensity of speckle dynamics. This picture shows the waterfall-charts of different SES types in vicinity of railroad ballast prism (trains, technological activities, etc.). Waterfall-charts (9 pieces) of this picture were obtained of different parts of FOS, on three frequency bands 4–10 Hz (left part), 20–40 Hz (central part), 60–150 Hz (right part). Table 3 contains the application results of used Classification Algorithm. When testing, there were two distances from FCCTV sensors to STE: 200 and 500 meters. Symbol α denotes a value of I Type error (false reject), and symbol β—a value of II Type error (false alarm). We can assume that the application results are not bad.

6 Conclusions

This section contains a description of modern approaches to construction and data processing in monitoring systems of super-extended objects. These approaches were applied in real monitoring systems and have confirmed their practical effectiveness.

References

1. Choi, K.N., Juarez, J.C., Taylor, H.F.: Distributed fiber optic pressure/seismic sensor for low-cost monitoring of long perimeters. In: Proceedings of SPIE 5090, Unattended Ground Sensor Technologies and Applications, pp. 134–141 (2003)
2. Korotaev, V., Denisov, V.M., Timofeev, A.V., Serikova, M.V.: Analysis of seismoacoustic activity based on using optical fiber classifier. In: Latin America Optics and Photonics

Conference, OSA Technical Digest (online) (Optical Society of America, 2014), paper LM4A.22 (2014)

3. Hampel, F.R, Ronchetti, E.M., Rousseeuw, P.J., Stahel, W.A.: Robust Statistics. The Approach Based on Influence Functions. Wiley, New York
4. Harrison, D.K., Hawthorn, I.E.: Amputation level viability in critical limb ischaemia: Setting new standards. Adv. Exp. Med. Biol. **566**, 325–332 (2005)
5. Huber, P.: Robust Statistics. Wiley, New York (1981)
6. Maronna, R., Yohai, D.M.: Robust Statistics. Theory and Methods. Wiley, New York (2006)
7. Shevlyakov, R., Smirnov, R.: Robust estimation of the correlation coefficient: an attempt of survey. Austrian J. Stat. **40**(1 & 2), 147–156 (2011)
8. Mei, Y.: Sequential change-point detection when unknown parameters are present in the pre-change distribution. Ann. Stat. **34**, 92–122 (2006)
9. Lai, T.L.: Sequential Change point detection in quality control and dynamical systems. J. R. Stat. Soc. Ser. B **57**, 613–658 (1995)
10. Elhoseiny, M., Bakry, A., Elgammal, A.: Multiclass object classification in video surveillance systems—experimental study. Proceedings of the 2013 IEEE Conference on Computer Vision and Pattern Recognition Workshops (CVPRW'13), pp. 788–793. IEEE Computer Society, Washington, DC, USA (2013)
11. Lowe, D.G.: Distinctive image features from scale-invariant key-points. IJCV **60**(2), 91–110 (2014)
12. Tan, M., Wang, L., Tsang, I.W.: Learning sparse svm for feature selection on very high dimensional datasets. ICML 1047–1054 (2010)
13. Dalal, N., Triggs, B.: Histograms of oriented gradients for human detection. In: IEEE Computer Society Conference on Computer Vision and Pattern Recognition (CVPR'05), vol. 1, pp. 886–893 (2005)
14. Laptev, I.: On space-time interest points. IJCV **64**(2–3), 107–123 (2005)
15. Wang, I., Klaser, A., Schmid, C.: Dense trajectories and motion boundary descriptors for action recognition. IJCV 103(1), 60–79 (2013)
16. Timofeev, A.V., Egorov, D.V.: Bimodal System for Surveillance of Super-extended Objects. Vol. 1, pp. 57–62. IWSSS-2014, At Bucharest (2014)
17. Gehler, V., Nowozin, S.: On feature combination for multiclass object classification. Int. Conf. Comput. Vision, IEEE 221–228 (2009)
18. Ratsch, G., Scholkopf, B., Smola, A.: Robust ensemble learning. In: Smola A.J, Bartlett P.L., Scholkopf B., Schuurmans D. (eds.) Advances in Large Margin Classifiers, pp. 208–222. MIT Press (1999)
19. Hears, M., Dumais, S., Platt, J., Scholkopf, B.: Support vector machines. IEEE Intell. Syst. **13**(4), 18–28 (1998)
20. Timofeev, A.V.: The guaranteed estimation of the Lipschitz classifier accuracy: confidence set approach. J. Korean Stat. Soc. **41**(1), 105–114 (2012)

Image Based Control of a Simple Mobile Robotic System

Dan Popescu, Loretta Ichim, Radu Fratila and Diana Gornea

Abstract This paper presents a mobile platform able to identify and track a spherical colored target and also to navigate upon preset trajectory, with avoidance of possible obstacles. The system can be configured for two scenarios. The first scenario is to move the platform upon a trajectory determined by the direction to a fixed point which is the center of a spherical colored object (the fixed target). The robotic system can avoid any obstacles occurring in the route. The second scenario is to identify and track a spherical colored object. In order to accomplish the above tasks the robot uses a simple navigation system based on a mono-camera and compass sensor. Except the mobile platform, the system has a central processing module for the analysis of received data and the elaboration of the control law. The task control is established by the aid of the extracted features from images. After obstacle avoidance the path correction is done both by compass and the camera information. The control algorithms are implemented on the robot microcontroller and the image processing algorithms are implemented on the PC (central module).

Keywords Image based control · Robotic system · Target detection · Obstacle avoidance

D. Popescu (✉) · L. Ichim · R. Fratila · D. Gornea
Faculty of Automatic Control and Computers, University Politehnica of Bucharest, Bucharest, Romania
e-mail: dan.popescu@upb.ro; dan_popescu_2002@yahoo.com

L. Ichim
e-mail: loretta.ichim@aii.pub.ro

R. Fratila
e-mail: radu_f90@yahoo.com

D. Gornea
e-mail: diana.gornea@yahoo.com

L. Ichim
"Stefan S. Nicolau" Institute of Virology, 285 Mihai Bravu, Bucharest, Romania

© Springer International Publishing Switzerland 2016
E. Pricop and G. Stamatescu (eds.), *Recent Advances in Systems Safety and Security*,
Studies in Systems, Decision and Control 62, DOI 10.1007/978-3-319-32525-5_7

117

1 Introduction

Object tracking [1–3] and obstacle avoidance [4] are two aspects of the research work in robotic domain which were studied in the last decade. Both are parts of the visual servoing, a concept widespread in modern robotics [5, 6].

Looking from a broader perspective, it would not be mistaken to say that mobile robots making use of video cameras for purposes of object tracking and obstacle avoidance can be widely met, from commercial appliances to research topics or even space missions. Precision, cost-effectiveness and flexibility in terms of environmental conditions and classes of recognized objects are, perhaps, some of the most important performance indicators when starting to develop a navigation robot.

In [3] a solution consisting in a low-cost single camera for spherical object tracking is proposed. In [2] the authors present a FPGA solution of a real time video system for spherical object tracking, also based on color information. They use the HSV color space decomposition, together with the Mahalanobis distance for precise location of the object in the scene, although it is pacing with high velocities, or in an irregularly lighted scene. For the tasks above mentioned two solutions were adopted in literature: one solution uses fixed cameras, and another uses mobile cameras.

Llamazares et al. [4] propose a solution to dynamic avoidance of obstacles based on Bayesian occupancy filter which is optimal with respect to both safety and consumption of energy. Their aim is to minimize the noise existing in the data coming from sensors, thus reducing the calculation complexity early from the acquisition phase. Since their approach involves bypassing mobile obstacles, a step for obstacle pace predicting is needed. This step is performed by optical tracking.

The most important information, video information is provided by a video camera which can be mounted on the moving platform (mobile camera) or on the outside, somewhere above the scene (fixed camera). The latter case can be represented by one or multiple cameras, which are used to supervise the entire environment of the robot in the field. In this way the location of the robot and/or the obstacles on its path can easily be mapped and represented in an absolute coordinate system. Nonetheless, the mobile camera approach provides the benefit of enhanced field of vision for particular detailed scenes that can only be taken on site.

Regarding the image processing for servoing or other tasks, three solutions can be adopted: (i) the data processing is made on a location outside the robot and the communication between the robot and processing unit is wireless [7]; (ii) the data processing is made inside of the robot platform [6]; (iii) a combination between the first two (i.e. partially outside the robot and partially inside the robot). The last solution is the base of our work.

In this paper we develop a low-priced and efficient solution of the problem of tracking a target (a defined object) simultaneously with possible avoidance of an obstacle. The solution is based on data aggregation and processing by two processing units: one on the robot platform (mobile) and another fix. The information is derived only from a video mono-camera and from a magnetic compass. Although we do not make use of other conventional sensors, like infrared beams or sonars, we obtained a satisfactory rate of success in both indoor and outdoor ambiances, with different levels of brightness.

The remaining of the paper is organized as follows: in Sect. 2 we present the system architecture and the proposed methods for control and image processing; Sect. 3 describes the details of hardware and software implementation; Sect. 4 presents some experimental results and, finally, in Sect. 5, we conclude and discuss strengths, weaknesses and feature development.

2 System Configuration

As mentioned above, the system was set up to accomplish two purposes: to track a target (object) and to avoid an obstacle that interposes in the moving direction. To accomplish this, the image processing and control of movement must be executed in real time.

To outline again, the main functional requirements of the robot are:

- maintaining a linear trajectory on a certain navigation path;
- automatic detection of obstacles and objects to be tracked in two ways: by color or by their contour, in different lighting conditions;
- automatic avoidance of obstacles interposed in the way, followed by a return to the original path;
- ability to automatically choose on which side the obstacle is avoided;
- ability to set direction and speed, and observe the road through the application's interface. This functionality can be extended to road mapping, providing that the images taken during robot's movement are stored.

As far as the methodology of addressing these requirements is concerned, we identified two approaches. The first one is to treat distinctively the issues of tracking a certain object and avoiding the obstacles that interfere on the way. In other words, this means to consider two separate tasks which can be alternated on demand.

On the other hand, the second approach handles the object tracking—obstacle avoidance duality as a whole. Initially, on system start-up, controlling the direction of the robot is done based on the image processing algorithm. The direction is set relatively to the tracked object, by permanently comparing the coordinates of the object's center of gravity to the coordinates of the center of the robot's field of vision. The moment when an obstacle is interposed between the robot and the object tracked, until that obstacle is bypassed, the direction is stored and further maintained using a compass. After surpassing the obstacle, the system's direction is set again according to the center of the tracked object. That is to say, immediately after passing the obstacle, there is an intermediate step of re-adjusting the forward direction: the compass is disconnected and the control loop on image processing feedback is resumed.

In the possible event that the object to be tracked is no longer found in robot's field of vision, the mobile platform will slowly rotate around its axis, using the lowest speed in order to give time for the image processing algorithm to identify again the object. There is the possibility to discover another object having the same criteria used for recognition as the original object. In this case, the robot will start following this new object. On the other hand, if no object matching the recognition criteria is found, after a complete rotation around itself, the robot will use again the forward direction provided by the compass, until the eventual detection of the object.

Video control of the system is based on a cost-effective solution, using one wireless camera for image acquisition. Also, the signal processing from and for mobile platform is done on a low-cost board, Arduino type. For the image processing and setting references and parameters of programs, a computing unit, laptop type (considered fixed for this process) is used. Between the two units (fixed and mobile) it is established a wireless communication. Through the program, object tracking, obstacle avoidance, or both tasks can be set. Architecture also offers the advantage of low power consumption.

The mobile platform has four bidirectional motors, two for each side of the platform, being controlled by an H Bridge. For ensuring and maintaining of the direction, a magnetic compass is used. This is located in the control loop with the H Bridge.

In addition to direction control loop, the system has loop control for speed and for target tracking (spherical object). In the latter case the control is based on the acquired image (servoing).

The system can determine the distance to the tracking object based on prior learning of the size and color of the object. The programmed references are, outside the object itself (color and shape): speed, minimum distance from object and the angle against North direction.

The robot has two operating phases: a learning one and an execution one. They are switched by the operator. In the first phase the system is learned with the object to be tracked further on. The system retains the color of the target, in terms of values for each spectral component, and the size of the target, in terms of object's diameter. This information is later used in the second phase to identify the target and to permanently compute the distance towards it, in order to maintain a certain distance between the followed and the follower.

The configuration of the system is presented in the block schema in Fig. 1 (Fig. 1a—the fixed and the mobile units, Fig. 1b—modules of the robot) and also in Fig. 2—intuitive architecture.

In the Fig. 1a, b we used the following abbreviations: AU—Arduino Uno, R—Robot (Mobile Platform), CU—Communication Unit, PC—Laptop, MU—Mobile Unit, FU—Fix Unit, C—Magnetic Compass, B—Battery, VC—Video Camera, HB —H Bridge, LM—Left Motors, RM—Right Motors and M1, M2, M3, M4—Motors.

In Fig. 2 we used the following components: (a) PC—Laptop, (b) Bluetooth Module, (c) Arduino Uno, (d) Compass, (e) H Bridge, (f) Mobile Base Structure and (g) WiFi Module.

Although the control subsystem is configured according to the application software, the acquisition of images from the camera, mounted on the mobile platform, is the same in both scenarios. As has been stated, the images from the video stream are acquired using a single video camera and this is a simple and

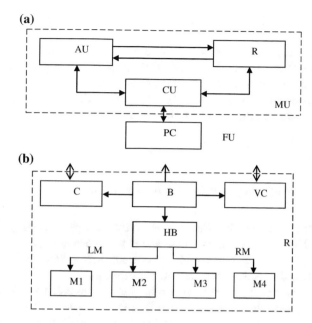

Fig. 1 System configuration: **a** general considerations; **b** robot details

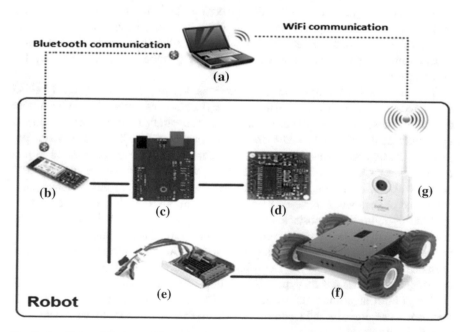

Fig. 2 Intuitive architecture of the system

low-cost solution, different from other options like stereo vision or using external video cameras.

The camera is a low-priced one, with a 640×480 pixels resolution with the following minimum requirements: low size, power consumption and cost, high sensitivity and wide viewing angle. One of the criteria based on which the video camera was selected is its wide viewing angle of $74.8°$. Therefore, selecting just a single camera will be satisfactory for this application, since it covers large portion of the ambient.

2.1 Data Processing for Obstacle Avoidance

The detection of possible obstacles is based on the image processing. The class of obstacles is defined in the learning phase and is based on chromatic information and the size. We proposed two algorithms for obstacle recognition: the first uses color information and the second uses contour information.

The first algorithm supposes that the road and the obstacle have different colors. In order to decrease the influence of the illumination, the HSV decomposition is used. The noise rejection is done by a median filter. After filtering, the thresholds on H and S establish if the obstacle color is present in the image. A region of similar pixels is considered an obstacle if it contains at least 5000 pixels.

The second algorithm requires edge detection. As in the first algorithm, image processing is done on application starting. Once acquired, images are converted to gray level images and are subjected to a median filter in order to eliminate any possible noises. After noise rejection, the edges (contours) are extracted by the Sobel operator.

This algorithm multiplies some contour lines, but this does not adversely affect the application, but rather is a positive factor because, the outline has to be further dilated. Because the Sobel operator introduces discontinued outlines, a dilatation of the resulting image is needed. After the outline is dilated, "the holes" are evened out to shape the objects (pixels that are part of the inside contour are considered as part of the object). Once the objects are identified, those objects that are directly connected with the image's edge are eliminated. In addition, as in the previous algorithm, the objects which have an area smaller than 5000 pixels will be removed.

After applying these obstacles recognition algorithms, either by detecting the color of the obstacle, or by detecting the outline of the obstacle, the computer transmits to the Arduino processing unit (on the robot), information about the presence of an obstacle and the algorithm to avoid it. The robot trajectory is on the left side or on the right side, depending of the horizontal coordinate of the centroid.

The control algorithms for maintaining direction of the robot and for obstacle avoidance are done by a PI controller. The reference for the displacement subsystem is the angle between the forward direction and the North direction. This information is done by the magnetic compass which is included in a closed-loop control. The increment is $1°$ and also the error is less than $1°$.

2.2 Data Processing for Target Detection and Tracking

We implemented both control algorithms to move the robot in a certain direction, following the target object, and control algorithms to return to the path desired after bypassing the obstacle.

The algorithm for target detection and tracking takes into account the following conditions which are mandatory since the learning phase:

- The object (target) is identified on color and shape (proposed spherical).
- The minimum and maximum dimensions (radius) are established;
- The horizontal coordinates of the centroid are calculated and memorized;
- The object radius at a known distance is determined.
- As in the precedent case (obstacles determination), the images are represented in HSV color space, from which H and S are considered.

In the learning phase, for the system calibration, the object color and object dimensions at the minimum and maximum distances (d_{min}, respectively, d_{max}) are considered. In the operating phase (measuring, recognition and tracking), after starting the application, from PC module the following steps are realized:

If there is a possible target object in the scene then:

(i) The horizontal coordinate of the centroid [8] is calculated i_{GO} to establish the moving direction to the target depending of the difference sign Δi (1):

$$\Delta i = i_G - i_{GO} \tag{1}$$

where i_G represent the horizontal coordinate of the center of the image. If $\Delta i > 0$ then the robot will be ordered to the left, if $\Delta i < 0$ then the robot will be ordered to the right, as it is explained in Fig. 3.

(ii) In order to give the command for tracking, the distance d from the mobile platform to the target object is determined. The distance is estimated by the

Fig. 3 Decision making for tracking phase

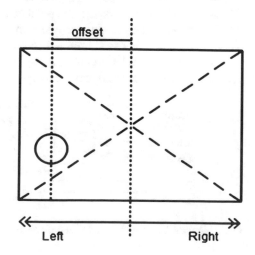

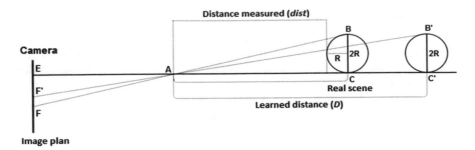

Fig. 4 Geometry used for finding the distance to the object

Eq. (2) and its elements are represented in Fig. 4 [3]. It must be noted that the distance must be kept in the established interval $[d_{min}, d_{max}]$.

$$d = \frac{F'E}{FE} \cdot D \qquad (2)$$

The control algorithms for tracking a target object is based on a servoing scheme, which operates with features extracted from images. The goal is to minimize the error (3):

$$e(t) = s(I(t)) - s^* \qquad (3)$$

where s(I(t)) represents the vector of features extracted from the image I at the moment t and s* represents the vector of the references feature. Obviously s(I(t)) is calculated from the image acquired by the video camera at the established frequency. The system needs an initial moment to fix a set point [6] and to identify the target object.

The control system is presented in Fig. 5. The significance of notations in Fig. 5 are: Uo ref—reference offset voltage; Ud ref—reference distance voltage; Uo—measured offset; Ud—measured distance; εo—offset error; εd—offset distance; CON—controller; M—motor; P—platform; C—camera; PI—image processing. The image processing module provides the voltages Ud and Uo which are compared with the references Ud ref and respectively Uo ref.

The output controller commands the motors so as to decrease the difference between the object centroid and the image center, and also to move the platform to

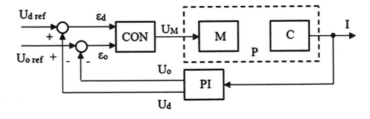

Fig. 5 Control loop based on offset and distance

the target. To control the speed and the direction, the two motors on one side of the platform are independently commanded against the motors on the other side. The control loop is a classical one and is implemented digitally on the Arduino.

2.3 Data Transmission

As the programs are created on PC and the image processing are also performed on PC in MATLAB, it is necessary a continuous communication between PC and Arduino and between PC and video camera.

Since the video stream is transmitted remotely via a wireless protocol, the application does not require a high performance camera because a high resolution broadcast video may hinder communication.

If, however, a high resolution video stream is chosen, the image processing unit may stop responding in real time, which will lead to losing control of the robot. On the other hand, if too low resolution images are chosen, there is a risk that objects will be no longer identified. In conclusion, there must be a compromise between image resolution and response time.

After performing a comparative analysis for available wireless communication protocols, for remote imaging a Wi-Fi device was used because it requires less transmission time compared to others. Sending commands to the robot from the processing unit does not imply a large flow of data transfer, and for that, a Bluetooth module was chosen because it has lower energy consumption.

The communication PC—camera is WiFi. The communication PC—Arduino is serial, by Bluetooth. There are two possibilities to separate the blocks of data, necessary to read continuously on Arduino:

- Introducing a separator character after each sent value and at the reception reading data until that character is met;
- Sending a fixed-length string and reading into blocks of that length.

3 System Implementation

Basically, the application consists in developing an intelligent robotic system whose main goal is detection and avoidance of obstacles found on its path. The robot was designed also to follow a navigation direction.

As it can be seen from the Fig. 1a, the system consists from two principal parts: a mobile one and a fix one. The fix subsystem is a PC which processes the images received from video camera through WiFi communication. The mobile subsystem contains the following elements:

- Video camera: Edimax IC-3115W;
- Magnetic compass: CMPS03;

- Processing unit: Arduino Uno board;
- Module for Bluetooth communication: Bluetooth Mate Silver
- H Bridge: Sabertooth 2 × 10 RC;
- Mobile platform: A4WD1 with 4 DC motors;
- Batteries for supply voltage: 12 V DC.

The system performs the following tasks:

- Set the navigation direction;
- Set the color, dimension and shape of the target object;
- Set the color, dimension and shape of the obstacle;
- Set the critical distances between robot and objects (target or obstacle);
- Establishing the speed of moving;
- Recognition of obstacle or target;
- Choosing the moment and angle for changing trajectory (for object tracking or obstacle avoidance);
- Periodic scanning of front side (the road);
- Maintaining the angle of basic navigation;
- Performing the algorithm of obstacle avoidance;
- Ensuring the motors command.

3.1 Software Development Environments

Essentially, MATLAB [9] is the programming language for implementing the proposed algorithms of control and data processing. MATLAB, Image Acquisition Toolbox and Image Processing Toolbox™ compose together a complete environment for the development of specific solutions in the image processing field. The acquisition toolbox allows for image acquisition frames to the same extend in which the camera or the computer can support high speed images.

Besides from acquiring images or videos, viewing data, developing algorithms and techniques for process analysis, we used MATLAB to create GUI interfaces as well. Our application has two user-interfaces, both for obstacle avoidance task and for tracking. The switching between the two GUI's is made by the operator. We have chosen this approach because the adjustments that need to be done prior to system starting (i.e. various calibrations) are slightly different for the two tasks.

Besides MATLAB, another software application named AMCap was used for interfacing the video camera for retrieving the video stream. We needed this extra layer in the middle of the video camera and MATLAB software because there was an incompatibility between the two. Therefore, it was necessary to create a software adapter so that MATLAB development environment to recognize the camera device and this was realized using freeware program AMCap.

The following pseudo-code lines describe the execution flow of the MATLAB application:

```
start
begin color_setup
  if choice == learn color
    capture image of target object;
    apply mean filter ;
    retrieve values  of the spectral components ;
  else if choice == use predefined color
    read values of the spectral components ;
  end;
return values of the spectral components;
end color_setup;
begin process_image
  convert image to HSV color space;
  decompose image to H,S,V components;
  for each spectral_component
    threshold : = color_setup.spectral_component;
    apply image binarization using threshold;
  end;
  compute minimum area value;
  if blob_area < minimum area
    ignore blob;
  end;
  enhance image;
  label remaining blobs;
  if no_of_blobs < 1
    display message: "No object found"
    break;
  end;
  for each blob
    extract features;
    if blob is circular
      compute x_offset;
      compute height;
    end;
  end;
return x_offset, height;
end process_image;
begin distance_calibration
  position target object to desired distance;
  capture image of target object in this position;
  set height of the target := process_image.height;
return setpoint_height;
end distance_calibration;
scan for available image acquisition devices;
for each device found
  display device_ID;
  if  device_ID ==  desired device
    start video;
  end;
end;
start serial communication
while video not stopped loop
  acquire frame;
  process_image(acquired_frame);
  serial communication of:
    process_image.x_offset,
    process_image.height,
    distance_calibration. setpoint_height
end loop;
stop
```

The Arduino IDE software environment is used for implementing the control algorithms and for establishing connections with the other components. The following Arduino libraries were used for setting up connections:

- Wire.h—used for establishing the link between Arduino board and the navigation module (the compass);
- SoftwareSerial.h—used for establishing communication with the Bluetooth module.

Further on, in the next pseudo-code lines we will present the algorithm implemented on Arduino:

```
start
start serial communication;
setpoint_direction := read from GUI;
setpoint_distance to the object := read from GUI;
while video not stopped loop
  while command not received
    wait;
  end;
  current_height of target := serialcom.read(height);
  setpoint_height of target :=
serialcom.read(setpoint_height);
  compute current_distance to the target using: cur-
rent_height and setpoint_height;
  current_offset of target := serialcom.read(x_offset);
  current_direction := read compass;
  switch task
  case: obstacle avoidance
    direction_err := setpoint_dir - current_dir;
    motors speed level := cruise_speed;
    if (center - current_offset)<0 then
      current_dir := current_dir-30degrees;
      delay;
      current_dir := current_dir + 60degrees;
      delay;
      current_dir := setpoint_dir;
      right_motor speed level +=  dir_err*PIDfactor;
    else
      current_dir := current_dir + 30degrees;
      delay;
      current_dir := current_dir - 60degrees;
      delay;
      current_dir := setpoint_dir;
      left_motor speed level += dir_err * PIDfactor;
    end;
  case: object tracking
    distance_err := current_distance - setpoint_distance;
    offset_err := center- current_offset;
    if distance_err > 0
       motors speed level := distance_err *P factor;
      else  motors speed level := 0;
      end;
      if (center- current_offset)<0
        right_motor speed level += abs(offset_err)*P;
      else  left_motor speed level += offset_err *Pfactor;
      end;
    end switch;
  end loop;
  stop
```

Fig. 6 Compass module CMPS03

3.2   Navigation Module—Compass

The module used for navigation is CMPS03 [10] (Fig. 6) which estimates the
difference between the forward direction and the North direction (horizontal
component). This compass uses the Philips KMZ51 magnetic field sensor, which is
sensitive enough to detect Earth's magnetic field.

However, a downside of using this kind of sensor is that any other outer magnetic field (for example, those produced by the motors of the mobile platform of the
robot) will interfere with the precision of the compass. In order to minimize this
issue, we used an aluminium shield made from tin foil.

In addition, the compass needs to be calibrated firstly. Calibration is done only
when the module is in operation and is necessary just once, afterwards data being
stored into its EEPROM memory so that it does not require to be performed each
time the module is turned on. The communication with the Arduino board is done
via an I2C interface.

3.3   Processing Units

Arduino Uno [11] is a development board using microcontroller ATmega328
(Fig. 7). It has 14 digital input/output pins (6 of which can be used as PWM
outputs), 6 analogue inputs, a 16 MHz internal clock, a USB connector, a power
jack, an ICSP head and a reset button. It contains everything necessary for the
functioning of the microcontroller; it will start to operate by simply connecting to a
computer via a USB cable or power to a power adapter or batteries.

Fig. 7 Arduino Uno board

Besides this embedded processing unit, we also used a laptop station, as part of the fixed processing unit responsible for the complex image processing [12]. The main configuration of this laptop station consists of: dual core processor with 2.1 GHz and 4 GB of RAM memory.

3.4 Microcontroller—Mobile Platform Connection

The connection between Arduino (microcontroller) and the platform's motors is done by an H bridge with two outputs connected one to the two left motors and two to the right motors (Fig. 8). The robot is equipped with off road tires, which make it suitable for outdoor environments as well. Because of its large wheels (120 mm in diameter) and high torque of the motors (gear ratio of 30:1), the robot has also very good stability which is important in case of sudden cornering due to obstacle detection.

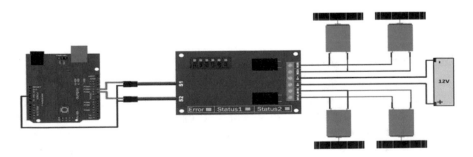

Fig. 8 Connection Arduino—H bridge—motors

4 Experiments and Tests for Different Algorithms

Testing experimental model was made in the following directions: the control of movement, identifying and avoiding obstacles, identifying and tracking objects of interest. For these purposes, it was implemented an interface (in MATLAB) between robot and operator, which is presented in Fig. 9.

The interface allows both analysis and interpretation of various images taken from the camera and set of operating parameters (by specific buttons): start communication, stop communication, speed range, angle of trajectory (compass), obstacle color, target color, minimum target dimension, different thresholds and trajectory for object avoidance.

The electrical diagram of a developing board, proposed as an alternative to the Arduino board, modeled using Proteus ISIS, is given in Fig. 10.

4.1 Experimental Results Concerning Motor Control

Robot speed v is established experimentally using different values for motor voltage and measuring the distance covered in 3 s. The connection between the supply voltage and the speed increment of the motor (also the distance covered during 3 s) is given in Table 1. When graphically represented, these measurements will compose an approximately linear dependence between motor voltage and the speed increment. It is the same with the dependence between the velocity of the robot and the speed increment.

The microcontroller generates pulse-width modulation signals (50 Hz, duty cycle between 5 and 10 %) to control the motors through the H bridge. An exemple of pulse-width modulation signal for H bridge is presented in Fig. 11. To be noted that the experimental values presented in Table 1 are dependent of the supply voltage of the H bridge. In other words, if the supply battery is running low, the velocity of the mobile platform will decrease as well. For the particular case, the measurements presented here were conducted at a supply voltage of 12 V for the H bridge.

4.2 Experimental Results Concerning Obstacle Avoidance

For experiments it was selected a red color for obstacle (algorithm 1 for obstacle detection) distance of 1 m and $30°$ angle to the left or right pass (Fig. 12), depending on the position of the center of weight of the object compared to the center of the image.

(a)

(b)

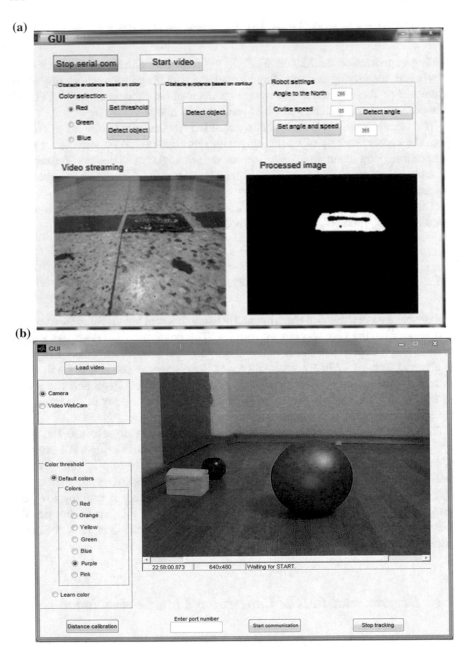

Fig. 9 GUI Interface of application **a** for obstacle avoidance and **b** for object tracking

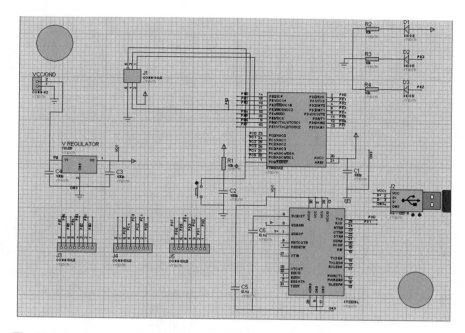

Fig. 10 Schematic diagram proposed as an alternative to the Arduino board

Table 1 Speed/Voltage dependence

Motor voltage (V)	Speed increment	Distance traveled in 3 s (mm)
0	0	0
0.5	1	105
1.0	2	230
1.5	3	365
2.0	4	495
2.5	5	615
3.0	6	720
3.5	7	845
4.0	8	960
4.5	9	1080
5.0	10	1200
5.5	11	1310
6.0	12	1430
6.5	13	1560
7.0	14	1670
7.5	15	1785
8.0	16	1900
8.5	17	2015
9.0	18	2130
9.5	19	2230
10	20	2335

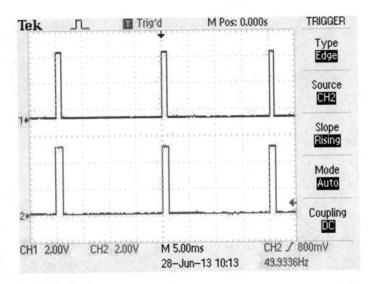

Fig. 11 Example of PWM signal for motor control having duty cycles of 5 and 7.2 %

Fig. 12 Trajectory around obstacle **a** *left side*; **b** *right side*

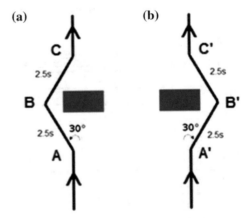

In the experiment a simple version with a small obstacle was considered. If the robot has reached the point A, where the obstacle has detected at a distance of 1 m, it will make a change in direction of 30° to the left that keeps 2.5 s for reaching in B point. In this point receives a displacement command in a direction of 60° to the right that keeps 2.5 s. After this it is considered that the robot has reached the point C located on the initial path, behind the obstacle. At this point the robot receives the displacement command on initial direction (avoided the obstacle).

4.3 Experimental Results on Tracking Targets (Objects)

The program for color selection and target recognition (object) is implemented in Matlab. The image representation was made in HSV color space, retaining only the components H and S, less sensitive to lighting and shadows.

For experiments, the green colour was chosen for the tracking object and the approximately round shape having the degree of eccentricity $e = \sqrt{1 - \frac{r_{min}}{R_{max}}}$ less than 0.6.

Video camera resolution was 680×480 pixels.

Comparing testing on static images with testing on video stream, image processing algorithm performance remained about the same, but the execution times increased (on average 1 s) due to wireless data transmission.

The minimum radius which an object can have is considered to be $r_{min} = 14$ pixels and the maximum radius is $R_{max} = 240$ pixels (half the image resolution). Table 2 shows the experimental results. The speed is 0 if the duty cycle is 75.

Table 2 Displacement parameters versus offset (pixels)

Offset (pixels)	Distance to the object (cm)	Test interpretation	Duty cycle for PMW1 (left motor)	Duty cycle for PMW2 (right motor)	Linear speed of the robot (cm/s)	Angular speed of the robot (°/s)
−320	20	Max right rotation	75	71	0	30
		Advancement 0				
+320	20	Max left rotation	71	75	0	30
		Advancement 0				
±5	340	Max advancement	55	55	72.4	0
		Rotation 0				
−320	340	Max advancement	55	51	72.4	30
		Max right rotation				
+320	340	Max advancement	51	55	72.4	30
		Max left rotation				
±5	20	Advancement 0	75	75	0	0
		Rotation 0				

These recordings represents extreme values (minimum and maximum) and medium values of the inputs of the tracking algorithm—left-right offset and distance to the object, both expressed in pixels. For each combination, the mobile platform's linear velocity (expressed as cm/s) and angular velocity (expressed as grade/s) have been measured.

It can be observed that, in situations with minimum and even medium inputs, the speed increments are relatively low, which means that the robot will move slowly. Both the advancement and rotation velocity of the platform will also be influenced by platform's mass (adding additional weights), but by supply voltage as well (discharging batteries).

5 Conclusions and Further Work

To sum up, in this paper we focused on building an intelligent robotic system that, based on image processing, manages to provide response to both obstacle avoidance and object tracking, two of the most extensively discussed topics in the literature. Our robotic system answers, at a given moment, to one of these two tasks, depending on how it is set up. Moreover, the implemented system is characterized by simplicity and efficiency.

One other particular achievement would be that our solution integrates different sensors and devices and uses data fusion from different sources to fulfill the allocated tasks with a cost constraint.

As far as the system's limitations are concerned, it should be mentioned that data processing including image is performed partially on mobile part and partially on fix part. For these reasons, due to communication delay, the system does not operate at high speed.

Further work will address the fully mobile architecture using FPGA implementation and the solution extension to a robot team.

Acknowledgement The work has been funded by the National Research Programme "STAR", project 71/2013: Multisensory robotic system for aerial monitoring of critical infrastructure systems—MUROS.

References

1. Guha, P., Mukerjee, A., Venkatesh K.S.: Efficient occlusion handling for multiple agent tracking by reasoning with surveillance event primitive. In: Proceedings 2nd Joint IEEE International Workshop on VS-PETS, pp. 49–55 (2005)
2. Popescu, D., Patirniche, D.: FPGA implementation of video processing-based algorithm for object tracking. U.P.B. Sci. Bull. Ser. C 72, 121–130 (2010)
3. Gornea, D., Popescu, D., Stamatescu, G., Fratila, R.: Mono-camera robotic system for tracking moving objects. In: 9th IEEE Conference on Industrial Electronics and Applications, Hangzhou, pp. 1820–1825 (2014)

4. Llamazares, A., Ivan, V., Molinos, E., Ocana, M., Vijayakumar, S.: Dynamic obstacle avoidance using bayesian occupancy filter and approximate inference. Sensors **13**, 2929–2944 (2013)
5. Hutchinson, S., Hager, G., Corke, P.: A tutorial on visual servo control. IEEE Trans. Robot. Autom. **12**, 651–670 (1996)
6. Cherubini, A., Chaumette, F., Oriolo, G.: A position-based visual servoing scheme for following paths with non holonomic mobile robots. In: International Conference on Intelligent Robots and Systems, pp. 1648–1654 (2008)
7. Taylor, S., Farinholt, K., Flynn, E., Figueiredo, E., Mascarenas, D., Moro, E., Park, G., Todd, M., Farrar, C.: A mobile-agent-based wireless sensing network for structural monitoring applications. Meas. Sci. Technol. **20**, 14 (2009)
8. Qahwaji, R., Green, R., Hines, E.: Applied Signal and Image Processing: Multidisciplinary Advancements. IGI Global Publishing (2011)
9. MATLAB: http://www.mathworks.com/products/matlab/
10. CMPS03—Compass Module: http://www.robot-electronics.co.uk/htm/cmps3tech.htm
11. Microcontroller Arduino Uno: http://arduino.cc/en/Main/arduinoBoardUno
12. Popescu, D., Ichim, L., Fratila, R., Gornea, D.: Reconfigurable robotic system based on mono-camera guidance. In: 6th International Conference on Electronics, Computers and Artificial Intelligence, pp. 45–50 (2014)

On Using a Cloud-Based Approach to Develop a Mobile Asset Management Solution

Dorel Nasui, Alexandra Cernian, Valentin Sgârciu
and Dorin Carstoiu

Abstract After the 9/11 events, stakeholders around the world have become concerned and looked for solutions to address the safety of the global transportation infrastructure. The Mobile Assets Management System (MAMS) presented in this paper focuses on improving transportation safety related aspects such as vehicle safety, cargo safety, passenger safety and driver safety. MAMS is a mobile asset management solution dedicated to locating and monitoring commercial fleets, offering fleet managers and dispatchers complete monitoring and control capabilities from an Internet-connected computer located anywhere in the world. The system is built on a cloud based secure, intelligent, interlinked and interactive infrastructure, namely the S4I I™ platform developed by SafeMobile. The implementation of the MAMS application is a reliable proof of the flexibility and the scalability of the S4I platform, truly creating a unique cloud-based integration system for wireless radio communication. We present the context and the benefits of the mobile asset management solution, the architecture and the main components of the system and we underline the advantages it provides to the global transportation infrastructure in terms of safety and security, which is a major concern nowadays.

D. Nasui (✉) · A. Cernian · V. Sgârciu · D. Carstoiu
Department of Automatic Control and Industrial Informatics,
University "Politehnica" of Bucharest, 313 Splaiul Independentei,
060042 Bucharest, Romania
e-mail: dorel@airadio.net

A. Cernian
e-mail: alexandra.cernian@aii.pub.ro

V. Sgârciu
e-mail: vsgarciu@aii.pub.ro

D. Carstoiu
e-mail: dorin.carstoiu@aii.pub.ro

© Springer International Publishing Switzerland 2016
E. Pricop and G. Stamatescu (eds.), *Recent Advances in Systems Safety and Security*,
Studies in Systems, Decision and Control 62, DOI 10.1007/978-3-319-32525-5_8

139

Keywords Mobile asset management · Cloud computing · S4I platform · Transportation infrastructure · Transportation safety · Remote safety terminal

1 Introduction

In the post-9/11 world, fleet managers, government officials, public safety personnel, and other stakeholders have become concerned about the potential for attacks involving the global transportation infrastructure. While recent events implore us to safeguard the transportation system from being used for violent acts, traffic accidents continue to provide the greatest threat to the safety of vehicles, cargo, and people. Therefore, stakeholders around the world have become concerned and started looking for solutions to address the safety of the global transportation system.

According to the National Highway Traffic Safety Administration [1], during 2013, 6 million police-reported traffic accidents in the United States injured 2.31 million people, took almost 33,000 lives, and caused $76.2 billion in property damage. Since the first recorded fatal traffic accident in 1899, crashes have killed more than 30 million people worldwide. By 2020, traffic accidents are expected to become the world's third leading cause of death [2].

The mobile assets management system (MAMS) presented hereby focuses on improving all aspects related to transportation safety: vehicle safety, cargo safety, passenger safety, driver safety, and it is based on a cloud-based secure, global transportation infrastructure that is intelligent, interlinked and interactive.

The main capabilities of the MAMS focus on identifying and averting potential threats, by letting the users know:

- where vehicles are at all times;
- when vehicles will arrive at their destinations;
- what routes drivers actually take;
- how in-vehicle systems are functioning;
- which vehicles are in need of maintenance;
- who the safest drivers are;
- how much each route is costing;
- whether the vehicle or container integrity has been compromised;
- if an emergency situation is developing.

Based on this powerful knowledge, the user can remotely activate or de-activate the vehicle or its internal systems. The Mobile Asset Management System (MAMS) offers fleet managers and dispatchers complete monitoring and control capabilities from an Internet-connected computer located anywhere in the world. In addition to enhancing the safety of vehicles, cargo, drivers and passengers, MAMS improves fleet management and reduces or eliminates losses from pilferage and unauthorized use. MAMS is compatible with a variety of wireless networks, so the existing investment in wireless equipment of a company is preserved.

2 Service Oriented Architectures in Cloud Computing

No doubt, the emergence and development of cloud computing is one of the most important technological revolutions in recent years. At the same time, this reality represents the natural evolution of a set of technologies, designed for utility computing. The fact is that for a large number of stakeholders, cloud computing plays a main role in the development of their technological strategies.

Cloud computing [3] is a new architectural concept of remote management of computing resources and data storage. The cloud technology has developed so much that now you simply create an account on Amazon or Yahoo in order to develop and launch applications in the Cloud. These are just two of the systems that are developed in the cloud, however, the services that can be developed can perform a wide variety of targets and can include almost any field (e.g. relational databases, applications, e-Commerce, CRM applications, ERP systems etc.). The cloud, as a concept, aims an efficient use of existing physical resources for maximum processing power and significantly enhanced efficacy compared to traditional processing methods [4]. They fully translate into virtualization of service dedicated workstations and servers—whether for data processing or storage, dynamic allocation of physical resources for these virtual machines based on individual needs at a time, the existence of reliable connections allowing Internet access and remote data processing. Figure 1 shows a graphical representation of the global vision of cloud computing according to [5].

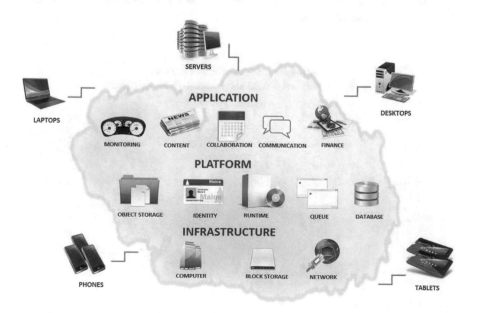

Fig. 1 A global vision of cloud computing [5]

Cloud computing is the convergence and evolution of several virtualization concepts, design of distributed grid and IT management applications, in order to allow a more flexible approach to implementing and scaling applications.

3   The Safemobile S4I Arhitecture

Safemobile S4I™ is a cloud-based Secure, Intelligent, Interlinked and Interactive application development platform that is able to collect and transmit information from a variety of data sources across multiple wireless communication systems [6].

The S4I platform enables comprehensive communications interoperability between different networks, products, devices and agencies by:

- Collecting information from various, heterogeneous sources, such as: audio, video and data terminals, sensors
- Transmitting and receiving information on multiple communication platforms
- Processing information both locally and in a secure, distributed environment
- Storing information both locally and in a secure, distributed environment
- Offering an extensive flexible, scalable, open interface for third-party applications
- Multiple language capabilities

Figure 2 presents an intuitive approach of the Safemobile S4I architecture.

The platform, having S4I at its core, can run on a variety of operating systems (Windows, Linux, OS2), being able to make functional connections between the equipment and software products developed by renowned companies (Oracle, Sun, IBM, Microsoft, Motorola etc.).

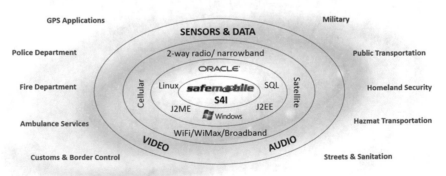

Fig. 2 S4I architecture

It is remarkable that the Applications Development Platform S4I can create configurations working with hybrid equipment (image, voice and data), remotely connected to wireless connections via the Internet or satellite, with drivers from different firms that use specific communication protocols difficult to reconcile in traditional conditions of use. The figure shows the possibilities of a platform with different wireless technologies, such as radio communications, cellular service, satellite or WiFi, Bluetooth and high-speed Broadband [7].

The platform can connect equipment that provides digital signals in various formats, from simple (analog sensors with analog-digital conversion and digital local, cellular or digital phone transmission by radio, tablets, pads) to the most complex which involve analysis, processing and transmission of video through different channels or VPNs, with sophisticated systems like TETRA or broadband.

The figure highlights, on the peripheral area, a number of potential applications of the S4I platform, such as: GPS applications, police department, fire department, ambulance services, Customs & Border Control, Street & Sanitation, MixMat Transportation, Homeland Security, public transport, military applications). These are just some of the possible applications, a wide variety of practical use of the platform being possible, according to user requirements.

What is not depicted in the picture above is the manner of resolving the hardware approach of the S4I platform [8–10]. All applications are developed for cloud computing, on a domain rented from Amazon.com which guarantees a chain of physical equipment: gateways, application servers, distributed database. In this way, a user can develop his own application by using built-in procedures stored in the platform, as well as his own contributions. The platform uses the most advanced and secure protection mechanisms, using SSL for each communication session and encryption/decryption of data packets.

The S4I applications development platform is an open architecture—open platform, which provides great versatility in developing solutions [11, 12]. The platform unifies information—private or public—in real-time nationwide, regardless of the manner of communication. Information is stored on secure servers and distributed to critical agencies and staff, regardless of their location around the world.

The fact that both applications and the entire quantity of results obtained by processing are kept in the cloud is a challenge for the real world, a desire imposed by the essence of cloud computing [13, 14]. Concerns about data security, integrity breaches that may be possible, as well as confidentiality issues have been resolved by way of conception and implementation of the S4I platform. All communications between equipment, devices, or other physical infrastructures and users (application developers, monitoring and surveillance personnel, other entities for maintenance of functional structures) is managed through the S4I platform based on cloud computing.

The S4I applications development platform is a configuration based on a service-oriented architecture, which has the merit of ensuring proper framework for interoperability.

Figure 3 shows a general structure of the SafeMobile architecture, with central focus on the S4I platform, which is responsible for integrating heterogeneous data

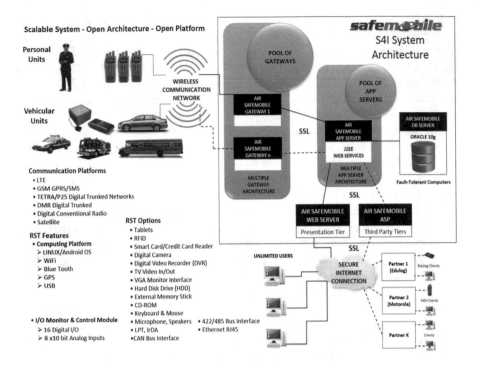

Fig. 3 S4I architecture

and processes, collected from various sources. The connection between input sources and the S4I platform is provided by a secure wireless communication network, which means that the S4I platform unites real-time information nation-wide. Information is collected through dedicated gateway servers.

4 Mobile Asset Management System Description and Architecture

The Mobile Asset Management System [15] is a location aware mobile asset management solution for operators of local, regional, and international commercial fleets, having the S4I platform, described in the previous section, at its core.

4.1 Mobile Asset Management System Main Features

The main features of the MAMS solution are the following:

Easy access. MAMS can be accessed via an Internet-connected PC located anywhere in the world.

Multiple fleets. To enable dispatchers to track the assets for which they are responsible, the fleet can be divided into individual sub-fleets requiring separate user names and passwords.

Current positions. The positions of all vehicles in the fleet are displayed on appropriately scaled maps. Maps can be customized to show locations of clients, depots, warehouses, garages, and other places of interest.

Historical position reports. MAMS allows to investigate where each of the vehicles has been over a period of time that the user selects—from the last 60 min to the past three months. MAMS provides four different types of historical position reports for this purpose.

Condition monitoring. Analog parameters, such as temperature or air pressure, can be displayed in table or graph format.

Messaging. When a compatible display device, such as a PalmOS handheld computer, is installed in the vehicle, drivers can send and receive text messages using MAMS. At the computer, the user can send a message to an individual driver, read all messages from an individual driver, or view all unread messages.

Poll. Just in case a vehicle hasn't sent a message in a while, the vehicle can always be polled for its current location.

Remote control. MAMS allows users to remotely control connected vehicle sub-systems. For example, the user can start the engine, lock a door, or flash lights and horns from anywhere in the world.

Geofencing. MAMS can be configured so that vehicles send a message every time they enter or leave a defined geographic area. MAMS provides the capability for users to remotely program as many as 244 distinct geofencing zones.

Configuration. Additional information, such as chassis manufacturer, model number, capacity, or any other vehicle characteristic can be associated with each vehicle. MAMS allows this information to be viewed and edited at any time.

Billing. The billing system allows online monitoring of the traffic related to each data terminal. MAMS allows users to view both a summary report and message detail.

Database. All messages are time-stamped and stored in a database that can be accessed by an Internet-connected computer (with appropriate access permission). The database format allows users to generate custom reports using a rich set of business analysis and data mining tools. The database engine currently deployed is Oracle 10 g.

Mobile Phone. Events, such as geofencing violations (meaning that a vehicle is leaving its authorized territory) can send alert messages to remote mobile phones. A mobile phone can also be used to configure vehicles if an Internet connection is not available or convenient.

Security. To make sure location data remains secure, all wireless transmissions are scrambled and/or encrypted, and remote access to MAMS is provided only via encrypted Internet connections.

Reliability/Availability. MAMS is designed to be highly reliable and available. Even in the event of a total system failure, total recovery can usually be accomplished in less than 60 s with no loss of data.

Administrative interface. MAMS also has a separate administrative interface for those who wish to customize their installation.

4.2 Mobile Asset Management System Components

An overview of the architectural approach of the mobile asset management system is depicted in Fig. 4.

But, how does the system work?

- Each vehicle has a tracking unit installed.
- Each tracking unit has built-in GPS and GSM modules.

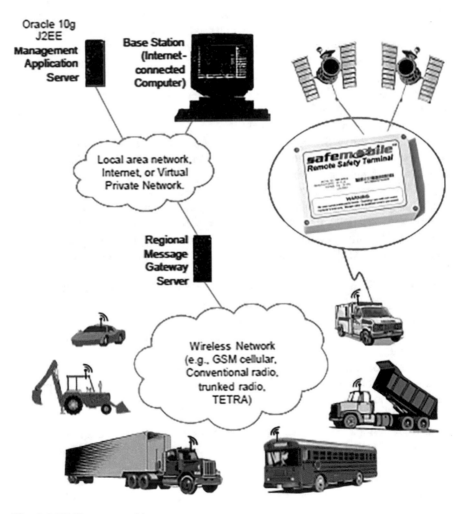

Fig. 4 MAMS system architecture

- The GPS receives messages from geo-stationary satellites and is able to compute the vehicle's physical position on the earth.
- The tracking unit monitors changes in the vehicle status and takes regular positional samples and stores them in its internal memory. The RST gathers and continuously records information like the vehicle's position, speed, and the status of its inputs.
- Once the memory becomes full or the unit is instructed to send, the stored information is sent to a dispatcher using the selected communication network.
- The Safemobile application server receives messages and displays the information via operator friendly electronic maps and database tables.
- The system has the capability to forward intelligent text messages to remote cell phones providing up-to-date information. The remote functionality also allows remote users to perform certain tasks while away from the dispatch center.

The MAMS solution consists of three main components, as depicted in Fig. 5.

Remote Safety Terminal (RST)

The RST is an Automatic Vehicle Location (AVL) transmitter and control unit that generates data about the vehicle's location and connected sub-systems. An RST is installed in every vehicle to be tracked. The RST calculates the vehicle's position every second and monitors the status of connected vehicle sub-systems, such as ignition, door sensors, and temperature sensors. The RST sends messages with location and status information via a wireless network to a SafeMobile message gateway.

RST models are available for transmission over a variety of wireless networks. The programmable RST has extremely flexible position and vehicle status reporting capabilities. Location and event reports can be transmitted periodically (e.g., every 10 min), upon defined events, or whenever a certain period has elapsed since the previous message.

The main features of RSTs are the following:

- 12 channel GPS receiver providing location accuracy typically less than 10 m
- ruggedized enclosure

Fig. 5 MAMS components

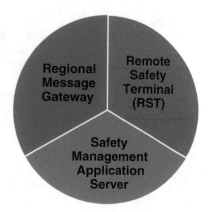

- dual processor architecture
- one communication and peripheral processor
- on application logic processor
- automatic reporting of sampled positions and inputs status change
- multiple positions (currently 20) compressed into a single message
- more than 600 positions stored
- geo fence alarming
- user defined event report mode
- polling request (on-demand location query)
- user selectable sampling and reporting rates and immediate reporting status
- sleep mode with user selectable wakeup-and-report mode
- text message communication to and from a PDA or text terminal
- user can send command to the remote units
- speed monitoring
- excess idling (stationary with ignition on)
- low power consumption

Figure 6 presents the strategic role of the RST within the mobile asset management solution.

Regional Message Gateway

A message gateway receives the wireless messages sent by the RSTs, decodes and re-packages them using a dedicated XML protocol, and sends them via a wired network to the Safety Management Application Server (SMAS). Regional Message Gateways are deployed as needed depending on the number of vehicles and their geographic distribution.

Once the message gateway receives the messages from the RSTs, it uses a dedicated XML protocol to decode and process them. Afterward, the messages are re-packaged and sent to the Safety Management Application Server (SMAS), using a wired network. The XML protocol has two components: the input protocol and the output protocol, described in what follows.

Fig. 6 The position of the RST within the MAMS

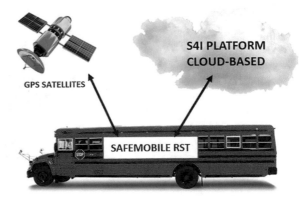

- **Input**

Command	Description
<AllPositionRequest/>	Request All Positions Ex: ?
<PollRequest vehicle_alias={alias}/>	Poll Request for vehicle Ex: @car
<VerboseModeRequest/>	Enter Verbose Mode Ex: V
<LogDumpRequest vehicle_alias={alias}/>	Log Dump Request for vehicle Ex: Lcar
<ChangeStatusRequest vehicle_alias={alias} status={1,2,3...8}/>	Change Status Request for vehicle Status may be {1,2,3...8} Ex: S2car
<OutputChangeRequest vehicle_alias={alias} output_number={1,2 ...8} state_value={F | T}/>	Output Change Request for vehicle Output number may be {1,2 ... 8} State Value may be{F | T} Ex: O2Fcar

- **Output**

Below is a sample of the XML protocol used for communication between the RST and the S4I platform.

```
<msg Vehicle="masina1467" Class="car" DateTime="11/01/07
08:07:04 AM" Contents="E0" EventId="10"
EventTxt="Timer 1" Fix="1"
Latitude="44.461960"    Longitude="26.102249"    Speed="0"
Bearing="1" Status="0"
Inputs="0000" Outputs="00" GeoData="00"/>
<msg Vehicle="masina1125" Class="car"
DateTime="11/01/07 08:06:18 AM" Contents="E0" EventId="10"
EventTxt="Timer 1" Fix="1"
Latitude="44.461960"    Longitude="26.102249"    Speed="0"
Bearing="1" Status="0"/>
```

Safety Management Application Server

To use MAMS, users connect to the Safety Management Application Server (SMAS) via a secure Internet connection. MAMS allows users to monitor and control their fleet anywhere, anytime, from any computer. The Safety Management Application Server is a fixed site with 2 primary function: receiving the messages from mobile units and store them into the secure Oracle 10 g database, and sent the information from Oracle database to user upon his request. These functions are implemented in 2 different modules (the J2EE application server and the web application) that work together through the Oracle database.

4.3 The User Interface

A rich, graphical interface provides clients with an intuitive way to view and control their fleets. This interface has the following features:

Current positions. The positions of all vehicles in the fleet are displayed on appropriately scaled maps. Maps can be customized to show locations of clients, depots, warehouses, garages, and other places of interest (Fig. 7).

Using the selection bar on the left side of the screen, the user can select only the vehicles of interest on the map.

Historical position reports. MAMS allows the user to investigate where each vehicle has been over a period of time—from the last 60 min to the past 3 months. MAMS provides four different types of historical position reports for this purpose.

Positions: The positions report depicts an icon of the vehicle at each position on the map at which the vehicle sent a message.

Route: The route report shows a directed line tracing the vehicle's progress.

Positions and route: The positions and route report plots both the directed line and the vehicle icons described above.

Table view: Instead of viewing on a map, historical vehicle positions may be displayed as a table. In addition to the absolute position, each entry in the table

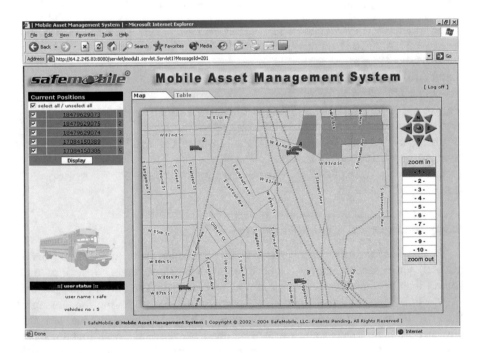

Fig. 7 Current positions

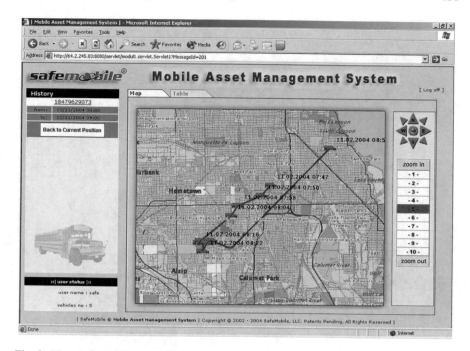

Fig. 8 Historical positions

includes the name of the nearest town center, the distance from the vehicle to that point, the position time-stamp (date/hour/minutes), the distance traveled over the reporting horizon, vehicle speed, and the status of digital inputs (Fig. 8).

Messaging. When a compatible display device is installed in the vehicle, drivers can send and receive text messages using MAMS. At the computer, you can send a message to individual driver, read all messages from an individual driver, or view all unread messages (Fig. 9).

Polling. Just in case a vehicle hasn't sent a message in a while, you can always poll a vehicle for its current location. To do so, simply click on the "send poll" button (Fig. 10).

Remote control. MAMS allows you to remotely control connected vehicle sub-systems. For example, you can start the engine, lock a door, or flash lights and horns from anywhere in the world. Just set the desired RST output level and click "set" (Fig. 11).

Geofencing. MAMS can be configured so that vehicles send a message every time they enter or leave a defined geographic area. MAMS provides the capability for users to remotely program as many as 244 distinct geofencing zones.

Configuration. Additional information, such as chassis manufacturer, model number, capacity, or any other vehicle characteristic can be associated with each vehicle. MAMS allows you to view or edit this information at any time.

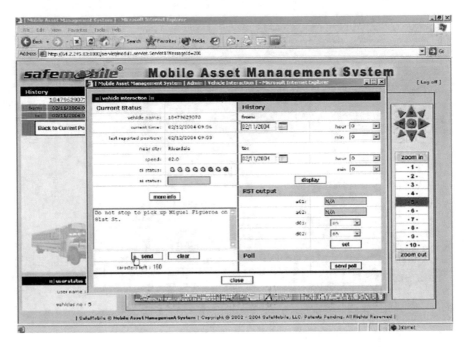

Fig. 9 Messaging

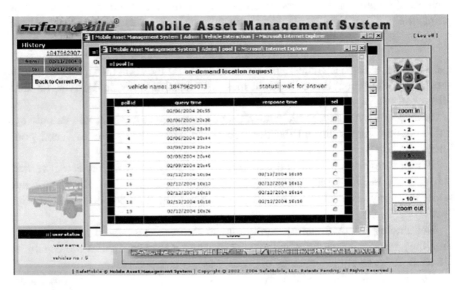

Fig. 10 Polling

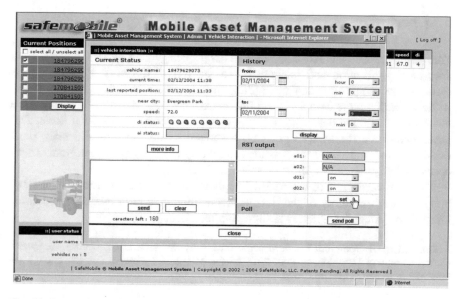

Fig. 11 Remote control

Billing. The optional billing system allows online monitoring of the traffic related to each data terminal. SafeMobile allows you to view both a summary report and message detail.

4.4 The System Administration Module

The System Administrator interface is used for configuration and maintenance of the MAMS system itself. Such tasks include registering vehicles into the system, creating new client accounts, and remotely programming the RST units, user administration, RST administration.

The System Administrator Interface also provides for RST administration. These operations allow a system administrator to address and configure individual RST units. Specific operations include the following: add/register new RST into the system, modify a registered RST's name, modify a registered RST's device attributes, send commands to an RST (extended command set), delete RST from database, show all registered RSTs that are not assigned to a fleet.

In certain areas, the System Administrator Interface may also be equipped with an optional billing system. With the billing system, the operator of a MAMS network can receive recurring revenues based on the number of wireless messages transmitted, and allow client bills to include message detail.

5 Conclusions

This paper presents a cloud-based mobile asset management system, developed as third-party application and integrated into the S4I platform. The application is dedicated to monitoring and controlling mobile assets anywhere in the world, without needing more than a PC and Internet connection. MAMS is a mobile asset management solution dedicated to locating and monitoring commercial fleets. This paper introduced the MAMS concept, described its architecture and main functionalities, and underlines the benefits brought by the solution to the global transportation infrastructure in terms of safety and security.

The implementation of the MAMS application is a reliable proof of the flexibility and the scalability of the S4I platform, truly creating a unique cloud-based integration system for wireless radio communication.

Acknowledgment The work has been partly funded by the Sectoral Operational Programme Human Resources Development 2007-2013 of the Ministry of European Funds through the Financial Agreement POSDRU/159/1.5/S/132395.

References

1. U.S. Department of Transportation, http://www.whitehouse.gov/omb/budget/fy2004/transportation.html. Last accessed 15 June 2015
2. Insurance Information Institute, http://www.iii.org/media/hottopics/insurance/test5/. Last accessed 16 June 2015
3. Raj, P.: Cloud Enterprise Architecture. Auerbach Publications (2012). ISBN 9781466502321
4. Joe, W.: Cloudonomics: The Business Value of Cloud Computing. Wiley, New York (2012). ISBN 9781118229965
5. Asociatia Nationala pentru Securitatea Sistemelor Informatice—GHID Securitatea in Cloud (2013). http://www.certro.eu/files/doc/775_20131030091057011764400_X.pdf
6. Nasui, D., Sgarciu, V., Cernian, A.: Cloud-based application development platform for secure, intelligent, interlinked and interactive infrastructure. In: Proceedings of IEEE 8th International Symposium on Applied Computational Intelligence and Informatics (SACI), 2013, pp. 473–476. ISBN 978-1-4673-6397-6
7. Balasa, F., Abuaesh, N., Gingu, C.V., Luican, I.I., Nasui, D.V.: Energy-aware scratch-pad memory partitioning for embedded systems. In: Proceedings of the 15th International Symposium on Quality Electronic Design (ISQED), 2014, pp. 653–659
8. Balasa, F., Abuaesh, N., Gingu, C.V., Nasui, D.V.: Leakage-aware scratch-pad memory banking for embedded multidimensional signal processing. In: Proceedings of IEEE International Conference on Acoustics, Speech and Signal Processing (ICASSP), 2014, pp. 5026–5030
9. Balasa, F., Luican, I.I., Zhu, H., Nasui, D.V.: Signal assignment model for the memory management of multidimensional signal processing applications. J. Signal Process. Syst. **63**(1), 51–65 (2011)
10. Balasa, F., Luican, I.I., Zhu, H., Nasui, D.V.: System-level exploration tool for energy-aware memory management in the design of multidimensional signal processing systems. In: ASP-DAC'09, Proceedings of the 2009 Asia and South Pacific Design Automation Conference, pp. 443–448. ISBN 978-1-4244-2748-2

11. Nasui, D., Rancea, I., Sgarciu, V., Dichiu, D., Oprea, B., Saru, D., Catana, I., Ceaparu, M., Tanase, C., Ene, C., Negulescu, C.: Supervising semi-autonomous mobile robots using safe-mobile wireless units. In: Proceedings of RAAD 2009—The 18th International Workshop on Robotics in Alpe-Adria-Danube Region, pp. 95–99, May 2009. ISBN 978-606-521-315-9
12. Nasui, D.V., Cosoi, A.C., Sgarciu, V., Rancea, I.: Using wireless monitoring for market research inverviewers. In: Annals of DAAAM & Proceedings, pp. 1153–1154, 2009
13. Balasa, F., Luican, I.I., Nasui, D.V.: High-quality data assignment to hierarchical memory organizations for multidimensional signal processing. In: Fifth Asia Symposium on Quality Electronic Design (ASQED 2013), pp. 89–96
14. Balasa, F., Luican, I.I., Zhu, H., Nasui, D.V.: Energy-aware memory allocation framework for embedded data-intensive signal processing applications. In: IEICE Transactions on Fundamentals of Electronics, Communications and Computer Sciences, vol. E92-A(12), pp. 3160–3168, 2009
15. Nasui, D., Cernian, A., Sgarciu, V., Carstoiu, D.: Cloud-based mobile asset management solution. In: Proceedings of ECAI 2014—International Conference, 6th edn. Electronics, Computers and Artificial Intelligence, 2014

Execution of an IEC61499 Application on a Remote Server

Oana Chenaru and Dan Popescu

Abstract This paper presents a solution for designing and executing IEC 61499 applications in a distributed web-based environment. This approach enables the separation of the input and output function blocks from the ones implementing the desired functionality that may require increased processing power. In this manner the functionality can be encapsulated in a generic, application independent structure and can be executed remotely using a web-based application described in this work. Standard communication interfaces are used for the information exchange of system modules and local devices. The system architecture, message format and details regarding the implementation are presented, as well as the results from evaluating the system functionality.

Keywords Remote execution · Distributed systems · Function block programming · Software integration

1 Introduction

Plant availability and resource efficiency from both materials and energy points of view are of utmost importance in any company. The current trend of next generation process control systems is to allow web access for the plant wide management

O. Chenaru (✉) · D. Popescu
Faculty of Automatic Control and Computer Science,
University POLITEHNICA, Bucharest, Romania
e-mail: oana.rohat@gmail.com

D. Popescu
e-mail: dan_popescu_2002@yahoo.com

O. Chenaru · D. Popescu
Department of Automatic Control and Industrial Informatics,
University "Politehnica" of Bucharest, 313 Splaiul Independentei,
060042 Bucharest, Romania

© Springer International Publishing Switzerland 2016
E. Pricop and G. Stamatescu (eds.), *Recent Advances in Systems Safety and Security*,
Studies in Systems, Decision and Control 62, DOI 10.1007/978-3-319-32525-5_9

and control strategies [1, 2]. This provides great accessibility, efficiency and flexibility for the process control engineers and gives the possibility of designing new applications like Smart Grids, remote executed systems or enterprise management systems.

Remote executed systems represent distributed systems where parts of an application are executed on remote servers, with the results being transmitted to a specific device using the Internet as the communication link. This means that a low resource controller can have access to the results from the execution of advanced algorithms without a resources upgrade.

The IEC 61499 standard [3] comes to support this trend and provides a methodology and standard communication objects [4] for implementing distributed systems. It promotes an application-centric approach in designing a process control system that focuses first on the functionality that needs to be implemented and last on the specific hardware allocation and integration.

This paper uses the design capabilities of the IEC 61499 standard to implement a distributed system that divides the algorithm execution, which will be done on an application server, from the commands transmitted from a plant controller to the field actuators. The novelty of this approach comes from the web integration of the remote execution.

The remote execution system is part of a web-based process control library that allows access to different algorithms from areas like process control, modeling and optimization, image processing, building management systems etc. This component allows a user to select a desired algorithm, configure the communication and input parameters, execute it on a remote server and send the results to a specific device identified by the IP address. The system is intended for use in non-critical applications based on scenarios like testing several algorithms to check which one suits best a specific implementation, running offline optimization algorithm for optimal input parameters identification, running image processing control methods, or control algorithms for slow processes.

The rest of this paper is organized as follows: Sect. 2 presents the data flow, the system modules and their interconnection. Section 3 describes modeling and implementation details of the remote execution system, how the data is encapsulated and processed at each level. Section 4 addresses the safety, security and transmission integrity aspects on the data link from the web interface to the remote device. In Sect. 5 a system prototype is evaluated. Conclusions are summarized in Sect. 6.

2 System Architecture

The remote execution system can be accessed by use of a web application designed as an algorithms library [4]. The system design follows the architecture illustrated in Fig. 1. This architecture shows how the decomposition of system functions, how

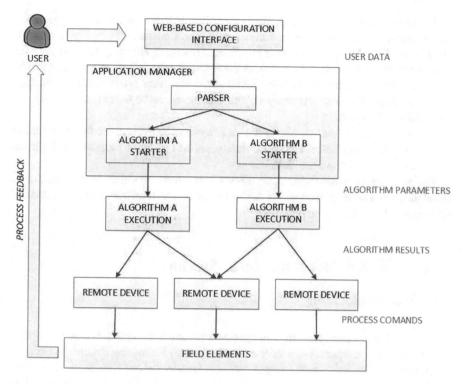

Fig. 1 System architecture

they were assigned to individual components and how the information is processed and transmitted at each level.

The user accesses the Web-based Configuration Interface, selects the desired algorithm and enters the parameters for establishing the communication with the remote device and the ones controlling the execution.

When the user requests the start of the algorithm execution, this data is forwarded to the application manager level. As each of these levels uses a different data format, a parser was implemented to analyze the user requests, extract de execution parameters and command the start or stop of an algorithm. The application manager uses multiple threads for handling different user requests, one for each algorithm instance. These threads are represented as the algorithm starter blocks.

The algorithm starter blocks are created dynamically, according to the user requests. A starter block servers a specific algorithm, identified by the ID parameter and for which the parser assigns specific communication arguments.

Each algorithm starter block can start or stop a single instance of a specific algorithm, according to the input parameters entered by the user. This execution can use only the web interface parameters, or can also receive plant information to refine the control program. The results obtained from the execution are sent to the remote device.

The remote device represents a plant controller, with its own processing unit and data acquisition modules. It uses a communication interface having the characteristics entered by the user in the web interface. This controller may be used only for commanding the specific field elements according to the values received from the algorithm execution, or can have its own logic that further processes the execution results. A controller with a processing unit can also receive execution results from more than one algorithm.

The user acts as an active observer that provides the system feedback, meaning that during the execution it can dynamically adjust the input parameters to obtain the desired reaction from the field device.

The presented system architecture allows increased scalability as the number of the algorithms and users supported depend only on the hardware characteristics of the server where the system resides.

3 Remote Web-Based Execution System

3.1 System Modeling

The IEC 61499 standard promotes an application-centric design that does not depend on the hardware characteristics of the future system. Following this methodology, several models for developing component-based model-driven application structures (MDA) are considered reference works in IEC 61499 system design. In [5] the author proposes a layered architecture called MIM (Model Integrated Mechatronics) that extends the MDA model by implementing abstract representations of reusable components that integrate the three engineering points of view of a mechatronic system: mechanical, electrical and software. In [6] the authors apply this methodology to distributed systems by developing application-independent automation components that can be embedded in existing control systems. In [7] the MDA model is mapped on the IEC 61499 specification by separation of the design process in three stages: PIM (Platform Independent Model), PDM (Platform Definition Model) and PSM (Platform Specific Model) to obtain hardware-independent model and thus achieve reusability of the code.

In the remote execution system design we applied the MDA model and obtained a Platform Independent Model and then applied it to Platform Definition Model of the distributed system as can be seen in Fig. 2 [4]. For this we considered a generic centralized application and separated the user input execution parameters from the algorithm implementing the desired functionality and from the commands resulted from the execution.

To obtain the PDM, the PIM was then split between three devices linked in a distributed network, each implementing a separate system component. Communication objects were added to ensure the data flow. This allows the geographical distribution of the system components in implementing the remote execution system.

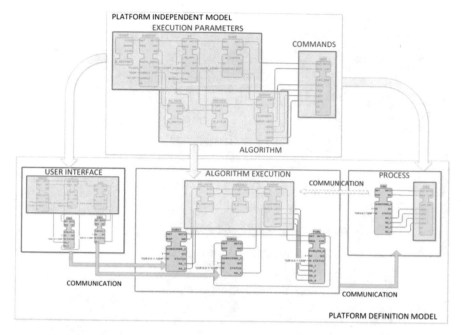

Fig. 2 Applying MDA for the design of the remote execution system

To integrate web connection capabilities, the user interface component was replaced by a web interface providing the same user interaction options. Because this web interface uses different technologies that could not implement the communication interface needed for the connection to the algorithm execution module, an additional application manager component was created. Its purpose is to ensure the integration between the different technologies in a way that does not affect the data flow and integrity.

This modeling approach is independent on the selected algorithm or on the process to which it is connected. This ensures a generic approach of the designed system that can be reused for any application.

3.2 Preparing the Application

The modeling approach described in the previous section was applied on a simple flasher application presented in [8] that is based on the FLASHER_TEST sample program (Fig. 3) from FBDK [8]. This program can be found in the src/doc folder. The FLASHER_TEST system configuration simulates a controller that receives as input a flashing frequency and a blinking mode. It controls four LEDs that will blink according to the user's input parameter values. This system represents the PIM implementing the desired functionality in a centralized approach.

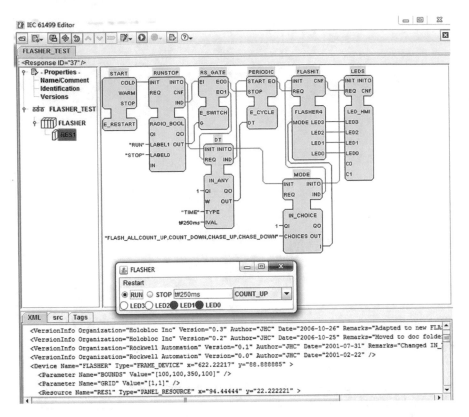

Fig. 3 FLASHER_TEST application in FBDK

The system configuration is developed according to the IEC 61499 standard as a network of function blocks. Their order of execution is controlled through the input events, while the links between the input and output data controls the information flow. The START function block controls the running of the system configuration. The RUNSTOP function block instantiates the radio buttons that control the execution for the following blocks though the RS_GATE function. This toggles between the start and stop of the execution. If the start sequence is activated, a periodic pulse is generated by the PERIODIC function block with a time frame specified by the user and received through the DT block. This PERIODIC block will enable a one-step execution of the FLASHIT block, according to the mode selected by the user. Five flashing modes are available:

- FLASH_ALL, where all LEDs are lit and then switched off at the time interval received as input in the DT block;
- COUNT_UP, which implements a binary counter from 0—all LEDS lightened, to 15—all LEDS switched off;
- COUNT_DONW, where the reverse counting from 15 to 0 is implemented;

- CHASE_UP, where only one LED is lit at a time, consecutively, from LED0 to LED3;
- CHASE_DOWN, which implements the consecutive lighting of a LED from LED3 to LED0.

For the implementation of this application in a remote execution environment, we must then build the equivalent application, distributed on three different devices, each with one resource. For this we must separate the user input blocks (RUNSTOP, DT and MODE) from the blocks implementing the desired functionality (RS_GATE, PERIODIC and FLASHIT) and from the blocks representing the output device (LEDS). To provide the data and event flow between the three sections, communication blocks must be added at the points where the splitting occurred. The communication blocks can be the publish/subscribe or the client/server function blocks available in FBDK in the net folder. Each pair must have a unique IP:port identifier. For testing purpose we chose the publish/subscribe function block that can be implemented in a local area network and an IP of 225.0.0.1 and ports 1024 and 1025 for the communication between the user input application and the functionality application, and port 1209 for the communication with the device. In a real application client/server blocks must be used for the communication with the remote device. The resulting applications are illustrated in Figs. 4, 5, and 6. A reevaluation of the functionality must be performed at this step, ensuring that the splitting and inter-application communication was done correctly.

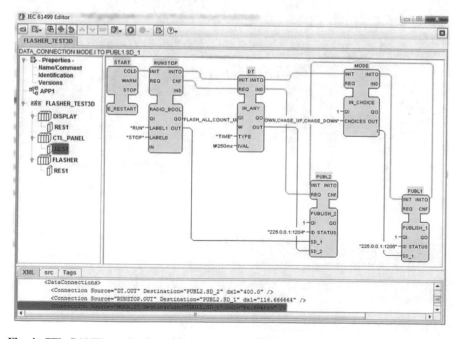

Fig. 4 CTL_PANEL application with the user input function blocks

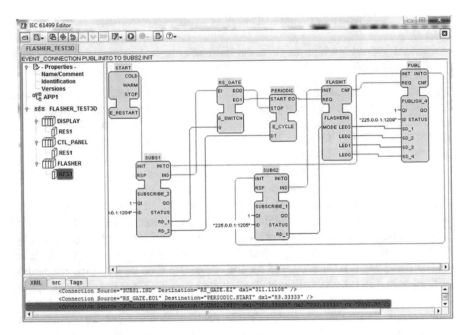

Fig. 5 FLASHER application with the function blocks implementing the desired functionality

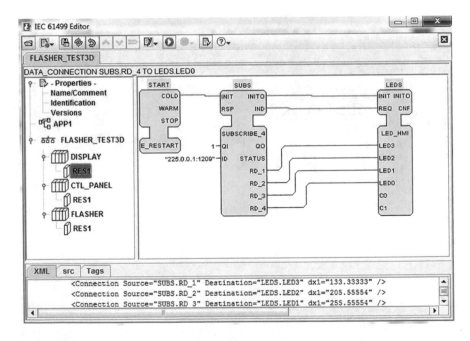

Fig. 6 DEVICE application for the control function blocks

The design of the distributed application should follow the recommendations presented in [8] to ensure that any further changes in the application are easily and correctly replicated in the function block instances.

From these three applications, only the FLASHER application will be stored in the process control library and will be executed on its server. The functionality of the DISPLAY application will be replaced by a web interface, allowing the user to enter the configuration parameters for the flasher execution. An Application Manager module was designed to retrieve this information and send it to the desired algorithm (FLASHER in our example). The DEVICE application can be implemented on a real controller or a remote PC.

3.3 Data Transmission and Processing

From the user interface to the remote device, data is analyzed at each level and different actions are executed according to each parameter's purpose, as can be seen in Fig. 7 [4]. The middle three components reside on the application server, which means data is transmitted between them using the localhost network. Even if the user can access the web interface from any geographical location, it is most probable that the commands for the execution are issued from the plant where the remote device resides as in this way the user can see the reaction of the plant to the input parameters.

3.4 The Web Interface

In the web interface illustrated in Fig. 8 the user selects the desired algorithm and this information is stored in an Alg. ID parameter, in this case "flash_test". It then completes information regarding communication information with the remote device in the Device IP address and Device TCP port fields (information stored in the IP:port parameter) and the input parameters needed for the algorithm execution in this case Frequency (Alg. Param parameter) and commands the start of the algorithm execution (Start/Stop parameter). This information is transmitted to the Web server by using standard GET/POST methods for retrieving a web page content.

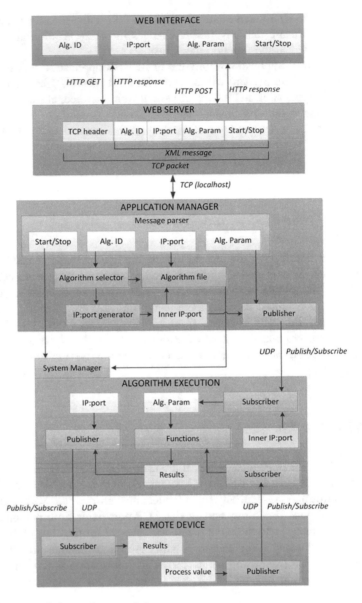

Fig. 7 Data transmission and encapsulation

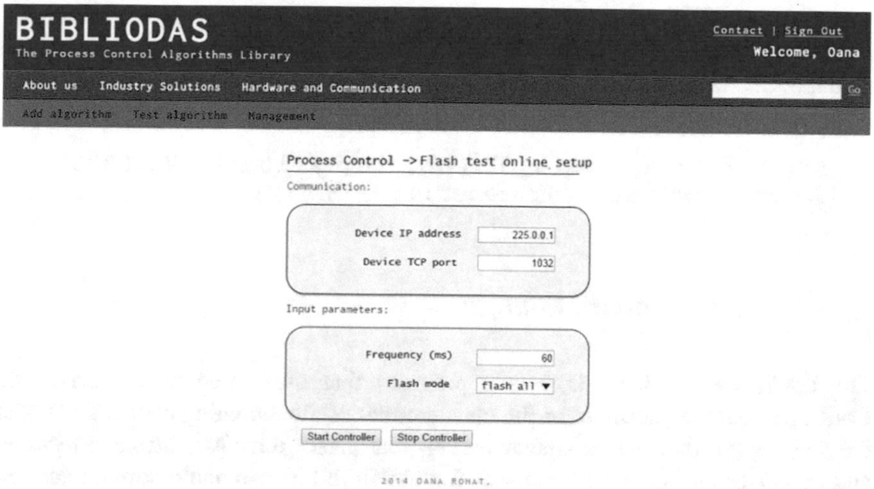

Fig. 8 Configuration of the remote execution in the web interface

3.5 The Web Server

At the web server, the parameters are encapsulated in an XML message and sent to the application manager using a standard TCP packet. For the flasher example, this is done using the following PHP commands:

```
$xml = new SimpleXMLElement("<xml/>");

$xml->addChild('sessionid', session_id());
$xml->addChild('username', $_SESSION['login_user']);
$xml->addChild('algorithmName', "flasher_test");
$xml->addChild('controllerIP', $controllerIP);
$xml->addChild('controllerPort', $controllerPort);
$xml->addChild('controllerAction', $controllercommand);
$xml->addChild('timebox', $timebox);
$xml->addChild('modebox', $modebox);

command_controller($controllerIP, $controllerPort,
$controllercommand, $xml);
```

Afterwards a new socket is opened and the message is send to the Java application. If the web server does not receive the confirmation from the application manager for the correct receipt of the message, it prompts a timeout error.

```
$fp = stream_socket_client("tcp://localhost:8080",
$errno, $errstr, $timeout);
```

3.6 The Application Manager

The application manager is a Java program that uses three main classes: the Controller class responsible for the management of the incoming connections from the web server, the WorkerRunnable class that parses the XML message received and generates an internal IP:port pair for use in the inter-module communication, and the System Manager class that prepares the algorithm for the execution according to the parameters entered by the user and starts this execution in FBDK.

The connection of a web interface instance to the application manager is done when pressing the Start Controller button. At that moment a new instance of **the Controller class** starts and a new execution thread is created. The server will open a new socket that will receive the data coming from the web server. This functionality is implemented as follows:

```
openServerSocket();
while (!isStopped()) {
  Socket clientSocket = null;
  try {
    clientSocket = this.serverSocket.accept();
  } catch (IOException e) {
    if (isStopped()) {
      System.out.println("Server Stopped.");
      return;
    }
    throw new RuntimeException("Error accepting client
connection", e);
  }
  new Thread(new WorkerRunnable(clientSocket)).start();
}
```

Once a new socket is received, the **WorkerRunnable class** is instantiated to parse the XML message received and store it in a string:

```
byte[] buf = new byte[1024];
StringBuilder stringBuilder = new StringBuilder();
InputStream in = clientSocket.getInputStream();

int i;
while ((i = in.read(buf)) != -1) {
   stringBuilder.append(new String(buf, 0, i));
}
```

After the string is parsed, the algorithm name and the remote device IP and port are extracted and a check is performed to ensure that this IP:port pair is not in use in another ongoing algorithm execution.

```
RemoteDevice rd = new RemoteDevice(remoteDeviceIP,
remoteDevicePort, algorithmName);

int index = runningDevices.indexOf(rd);
if (index != -1) {
    RemoteDevice current = runningDevices.get(index);

return "IP " + remoteDeviceIP + " and TCP port " +
remoteDevicePort + " is already in use by " +
    "algorithm " + current.AlgName;
    }
runningDevices.add(rd);
```

Afterwards the start/stopped commands is processed. At a stop request, the application parses the algorithm name and the corresponding instance related to that session ID is closed. If the user issued a start command the application will then command the IP:port generator to provide a unique set of multicast IP:port parameters that will be recorded in the Inner IP:port variable. This variable will be used for the transmission of the Alg. Param values to the algorithm execution module. A hashmap is created to store all the information associated between an algorithm execution and the corresponding device. All the information received from the web server will be stored here and forwarded to the **SystemManager class.**

```
    String sessionId = xmlFields.get(XMLKeys.SessionId);
    String algName =
xmlFields.get(XMLKeys.algorithmNameKey);
    String plantIP = xmlFields.get(XMLKeys.PlantIPKey);
    String plantPort =
xmlFields.get(XMLKeys.PlantPortKey);

    lastGeneratedPort++;
    if (lastGeneratedPort > 65535) {
      lastGeneratedPort = 1024;
      lastGeneratedIP.add(BigInteger.valueOf(1));
    }
    Integer algPort = lastGeneratedPort;
    byte[] bytes = lastGeneratedIP.toByteArray();

    byte[] fourbytes = new byte[4];
    for (int i = 0; i < 4; i++) {
      fourbytes[i] = bytes[i+1];
    }
    InetAddress address =
InetAddress.getByAddress(fourbytes);

    AlgorithmDevice algorithmDevice = new
AlgorithmDevice(algName);
    algorithmDevice.setAlgIP(address.toString());
    algorithmDevice.setAlgPort(algPort);
    algorithmDevice.setPlantIP(plantIP);
    algorithmDevice.setPlantPort(plantPort);
    perSessionInfo.put(sessionId, algorithmDevice);

    new Thread(new SystemManager(sessionId, xmlFields,
algorithmDevice)).start();
```

The SystemManager class receives as input the algorithm name, creates a copy of the algorithm file and renames it using the session ID. The generic macros from the initial algorithm file are replaced with the corresponding values of this instance.

The communication between the application manager and the algorithm execution modules, as well as the one between the algorithm execution and the remote device modules was implemented using the Publish/Subscribe mechanism. This is a connectionless, point-to-multipoint, unidirectional communication pattern that can be used for local communication in a distributed network [8, 10].

The Inner IP:port and the IP:port parameters will be written as input values of the Subscriber and Publisher function blocks of the selected algorithm, directly in the algorithm file. This will ensure that the communication will be established before the actual execution of the algorithm. Also, the Inner IP:port parameter will be used to configure the communication settings of the publisher from the application manager.

```
   String algorithmSystemFileName = "uploads/algorithms/"
+ algorithmName;
   Path srcFile = Paths.get(algorithmSystemFileName +
".sys");
   Path dstFile = Paths.get(algorithmSystemFileName + "_"
+ SessionId + ".sys");

    String newIPAndPort = new String(algIP + ":" +
algPort);

   Charset charset = StandardCharsets.UTF_8;
   content = new String(Files.readAllBytes(srcFile),
charset);
   content = content.replaceAll("__WEB_IP_AND_PORT__",
newIPAndPort);

   newIPAndPort = new String(AlgorithmDevice.plantIP +
":" + AlgorithmDevice.plantPort);
   content =
content.replaceAll("__BIBLIODAS_GENERATED_IP_AND_PORT__",
newIPAndPort);

  /* Replacing algorithm parameters */
  Integer i = 1;
  for (String key : FieldsMapping.keySet()) {
    String algParamValue = FieldsMapping.get(key);
   content = content.replaceAll("ALG_PARAM_" + i,
algParamValue);
      i++;
    }
   Files.write(dstFile, content.getBytes(charset));
   startSystem(dstFile);
  }
```

3.7 The Algorithm Execution Module

After these steps, the algorithm (now stored in the dstFile), is ready for the execution. The algorithm is executed using the FBDK tool [8]. To automate the execution of the algorithm by use of the web interface instead of the FBDK user interface, the application manager connects directly to the SystemManager, a tool behind FBDK that controls its execution. This way of interaction also allows us to have several instances of the same algorithm, which could not have been possible in a normal execution because of network conflicts.

```
String SystemManagerClass = "fb.rt.tools.SystemManager";
Class[] argTypes = new Class[1];
argTypes[0] = String[].class;

Method mainMethod =
Class.forName(SystemManagerClass).getDeclaredMethod("main
", argTypes);
String[] mainArgs = new String[1];

mainArgs[0] = file.toString();
mainMethod.invoke(null, (Object) mainArgs);
```

The SystemManager starts the execution of the algorithm file received as input. Through the Subscriber block the algorithm receives the Alg. Param parameters entered by the user in the web interface. The algorithm functions are processed and executed based in the received input parameters. The results are sent to the publisher block, configured with the communication details of the remote device. They are sent to the corresponding subscriber block from the remote device using UDP multicast packets.

If the algorithm needs process data for its execution, then an additional publish/subscribe pair must be implemented between the remote device and the algorithm execution modules. It will use the same IP from the IP:port parameter and a different port entered by the user in the web interface.

At any point when the algorithm is running, the user can send the stop parameter by pressing the corresponding button in the web interface, information that follows the same link until it reaches the SystemManager which stops the algorithm execution.

3.8 The Remote Device

Data is received by the remote device through a subscriber or client communication interface, pair to the one implemented in the remotely executed algorithm. In this testing configuration we used a subscribe interface, with the IP and port similar to the ones entered by the user in the web interface. This interface will be implemented on the plant device by the user, in a IEC 61499 compliant environment.

4 Aspects Regarding Communication Quality

Remote control rises important issues when it comes to the safety, security and integrity of the transmitted information. The security of the communication refers to the identification of an application vulnerabilities from the external factors (for example unauthorized users) point of view that may affect the integrity of the transmitted data [11]. The main measures for security management refer to access control and firewall installation. Safety refers to possible faults that may occur at the device level, faults that may interfere with the physical data transmission links. The main measures that can be applied to prevent such situations are ensuring the redundancy of critical devices and communication links, defining risk and hazard handling strategies and adding specialized emergency intervention devices (like ESD—Emergency Shut Down controllers).

Quality communication aspects must be analyzed separately at each entity: the web application, the application manager and the plant, and also for each communication link between these entities.

Web application access is controlled by the requirement of authentication credentials. The connection to the web server from the application manager is secured by enabling the HTTP protocol to run over the SSL—Secure Sockets Layer (HTTPS).

At the transport layer of the communication link between the web browser and the web server the connection uses the TCP—Transmission Control Protocol. This is a reliable transport protocol that uses sequence and acknowledgement numbers to ensure data integrity and its correct delivery. In case of packet loss, the TCP protocol has a retransmission mechanism that will guarantee the correct order of the sent data.

Security at the application manager level is ensured by a firewall that allows only TCP access on ports 80 (for HTTP) and 443 (for HTTPS). Also, antivirus and antispyware software with the latest security patches and updates ensure a proper security control at this level.

Communication between the application manager server and the remote device follows the publish/subscribe mechanism. In its basic implementation from FBDK [8], as is also presented in the IEC 61499 standard definition, a publisher uses UDP —User Datagram Protocol multicast frames so that one transmitted packet can reach multiple subscribers that are members of the same multicast group [4]. UDP is considered an unreliable protocol as it has no recovery mechanism in case of packet loss. UDP is suitable only for local area networks (LAN) where the transmission collision domain is limited and packet loss is minimum. UDP packets reside only on the local network because if the multicast group is not a general purpose group then the publisher message will be blocked by the first encountered router.

To overcome the data integrity and transmission limitations imposed by UDP, we chose to implement a secure VPN (Virtual Private Network) connection between the web server and the remote device from the plant. This represents a common solution used by process control engineers to ensure the quality of the data transmitted over the Internet [11–13]. A VPN creates a new virtual network connection on the application server which is in the same subnet as the remote device and all the traffic between the two end points is encapsulated or encrypted over the public network. Because VPN offers end-to-end connection, a tunnel is created between the two and the UDP multicast packets will travel over the public network.

A VPN tunnel can be configured with either a UDP or TCP transmission. By default, VPN will be configure as a UDP tunnel, meaning each transmitted packet will be encapsulated as a UDP packet. Even thaw UDP is not very reliable, the VPN solution works because traffic like HTTP, instant messaging or FTP rely on TCP as the transport protocol and even if a packet in the VPN is lost because of the UDP traffic handling, the inner TCP protocol will handle the packet loss. For our system, since the publish/subscribe mechanism uses UDP, we must change the default VPN settings to TCP so that it will handle the packet loss. This way we have the guarantee that packets sent by the publisher reach all subscribers, thus ensuring the data integrity.

From the security point of view, VPN requires the authentication of the client node with a dedicated user and password. This prevents unauthorized users to access or modify the transmitted information. Also, transmitted data can be encrypted based on the IPSec—Internet Protocol Security or the GRE—Generic Routing Encapsulation methods.

For the implementation of the VPN network, the remote device must be connected to a router with Internet access, using a public IP, which acts as a VPN server. It will dictate the user and password that the application manager will use to establish a connection and transmit the algorithm execution results.

The presented remote execution system is intended for use in non-critical applications. While the VPN-based solution minimizes data loss and ensures an acceptable level of integrity, additional safety measures must be considered when connecting to a high risk plant. That is because events like network signal loss during an algorithm execution can affect the process dynamics and can lead to its instability. Handling such situations implies developing a specific risk analysis, implementing algorithm execution error management mechanisms and delay handling methods [14].

A more complex solution that would also provide increased scalability represents integrating a Cloud solution specialized in providing services dedicated for process control applications like [15]. This offers secure connections, high connection speed and data transfer rates and it even supports data exchange through mechanisms like publish/subscribe or OPC.

5 System Evaluation

The system reliability was tested by developing a simple network configuration of two computers from different ISPs (Internet Service Providers), one implementing the application manager and one simulating a remote device. Both computers have Ethernet cards with the link speed of 100 Mbps. Both computers use the Windows 7 operating system. We analyzed the behavior of the system in different abnormal situations.

5.1   Network Signal Loss

A network signal loss was simulated by unplugging the network cable of the application server during the algorithm execution. This leads to the closing of the VPN connection and no data is transmitted to the remote device.

After the cable was plugged in, the VPN connection was not reestablished automatically. This is because in case of network signal loss Windows closes all the sockets associated with a network adapter, including the VPN connection. On the other hand, in Linux the sockets associated with a network adapter will remain open and the packets that are sent by any application will be buffered in the operating system until the network adapter will be up.

5.2   Network Congestion

A network congestion was simulated by downloading a large file (a movie file) over the established VPN connection. We did not use any quality of service mechanism to prioritize the traffic from application Manager. While the file transfer was running, we tested the algorithm start and stop commands and did not observe a behavior that is different from normal execution. During this period we also observed how changes in the algorithm input parameters lead to the corresponding adjustment of the execution results and there were no noticeable delays.

5.3   Application Manager Fail

To determine the system recovery after failure, we tried to see what happens is we stop the Application Manager. When the Application Manager receives a Start command for the execution of a new algorithm instance, it creates a new Java thread and give control to SystemManager class for running the algorithm. Since all the threads created by the Application Manager are part of the same process, when closing the Application Manager window, all the other threads will be closed as well. This means all current algorithm executions are stopped. This leads all the plant controllers in an inconsistent state as they cannot be controlled anymore and there is no notification from the Application Manager that the algorithm execution has stopped.

An improvement to this problem is that when the Application Manager is restarted, a check of the all algorithm status is performed based on the information from the system database. All running algorithms should be reset.

6   Conclusion

This paper presents a web based remote execution system that uses the capabilities of the IEC 61499 standard for implementing distributed applications. The presented work extends the application domain of the IEC 61499 standard to web based applications by providing mechanisms for accessing the web interface data and for dynamic connectivity to the input parameters of a function block using the FBDK environment. In addition to this, the authors identified a solution for controlling the execution of a FBDK application from a Java module using the SystemManager application.

By applying a MDA model in the design of the remote execution system we obtained a generic architecture that can be applied on any algorithm from the process control library and is able to connect to any IEC 61499-compliant application. This ensures the reusability not only at the algorithm representation level, but also at the application level.

The system implements safety, security and data integrity mechanisms that ensure a standard level of transmission quality. Future work will include developing advanced algorithms for the verification of the integrity of the transmitted information, delay handling [14] and enhanced communication protocols for real-time process control [16]. This will extend the domain of use for the presented system to critical or with hard real-time constraints.

References

1. Florea, G., Dobrescu, R., Popescu, D., Ocheana, L., Rohat, O.: PH Center a step to the next generation of process control architecture. In: Proceedings of the 11th WSEAS International Conference on Systems Theory and Scientific Computation, Aug 2011
2. Ferreira, P., Reyes, V., Maestre, J.: A web-based integration procedure for the development of reconfigurable robotic work-cells. Int. J. Adv. Rob. Syst. (Jan 2013)
3. Lewis, R.W.: Modelling control systems using IEC 61499: applying function blocks to distributed systems. In: Control Engineering Series 59, IEE, U.K. (2001)
4. Rohat, O., Popescu, D.: Remote web-based execution of IEC 61499 function blocks. 2nd International Workshop on Systems Safety & Security IWSSS, Oct 2014
5. Vyatkin, V.: IEC 61499 as enabler of distributed and intelligent automation: state of the art review. IEEE Trans. Industr. Inf. 7(4), 768–781 (2011)
6. Thramboulidis, K.: Model-integrated mechatronics—toward a new paradigm in the development of manufacturing systems. IEEE Trans. Industr. Inf. 1(1), 54–61 (2005)
7. Cengic, G., Ljungkrantz O., Akesson, K.: A framework for component based distributed control software development using IEC 61499. IEEE conference on emerging technologies and factory automation 2006, pp. 782–789, 2006
8. Zoitl, A., Vyatkin, V.: IEC 61499 architecture for distributed automation: the 'glas half full' view. IEEE Ind. Electron. Mag. 3(4), 7–23 (2009)
9. Vyatkin, V.: IEC 61499 Function Blocks for Embedded and Distributed Control Systems Design. ISA Publishing, USA (2012)
10. FBDK: Resources for the new generation of automation and control. Function Block Development Kit (FBDK). www.holobloc.com/

11. Andren, F., Strasser, T., Zoitl, A., Hegny, I.: A Reconfigurable Communication Gateway for Distributed Embedded Control Systems. In: Proceedings of 38th Annual Conference of the IEEE Industrial Electronics Society, 25–28 October 2012, Montreal, Canada
12. Yang, L., Yang, S.H.: A framework of security and safety checking for internet-based control systems. Int. J. Inf. Comput. Secur. 1(1/2) (2007)
13. JUNIPER Networks. Architecture for secure scada and distributed control system networks. White paper, 2010
14. Jadhav, M., Gidveer, G.: Internet based remote monitoring and control system. Int. J. Adv. Eng. Technol. (March 2012)
15. Tipsuwan, Y., Chow, M.Y.: Control methodologies in networked control systems. Control Eng. Pract. 11 (2003)
16. Skkynet. http://skkynet.com/solutions/industrial/
17. Esposito, C., Platania, M., Beraldi, R.: Reliable and timely event notification for publish/subscribe services over the internet. IEEE. Trans. Networking 22(1) (2014)

Evolution of Legal Issues of Honeynets

Pavol Sokol and Ján Host

Abstract Honeynets are unconventional security tools to study techniques, methods, and goals of attackers. It is very important to consider issues affecting the deployment and usage of these security tools. This paper discusses the legal issues of honeynets taking into account their evolution. Paper focuses on legal issues of core elements of honeynets, namely data control, data capture, data collection and data analysis. This paper also draws attention to the issues pertaining privacy, liability, jurisdiction, applicable law and digital evidence. The analysis of legal issues is based on the EU law.

Keywords Honeypot · Honeynet · Legal issues · The EU law · Privacy · Liability

1 Introduction

Several tools, methods and techniques aimed at protecting and securing communication between two or more network services have been successfully used for years now. Recently, they seem to have become ineffective against new and advanced security threats, therefore it is absolutely necessary to change and modernize techniques and tools that are used to protect networks from attackers. The honeypot and honeynet principle represents a different approach, which can defend given systems more effectively.

A **honeypot** is a "security resource whose value lies in being probed, attacked, or compromised" [1]. Honeypots are mainly categorized by the level of interaction, purpose, role and deployment. In this paper we will focus on the categorization by

P. Sokol (✉) · J. Host
Institute of Computer Science, Faculty of Science,
Pavol Jozef Šafárik University in Košice, Jesenná 5, 04001 Košice, Slovakia
e-mail: pavol.sokol@upjs.sk

J. Host
e-mail: jan.host@student.upjs.sk

© Springer International Publishing Switzerland 2016
E. Pricop and G. Stamatescu (eds.), *Recent Advances in Systems Safety and Security*,
Studies in Systems, Decision and Control 62, DOI 10.1007/978-3-319-32525-5_10

level of interaction, specifically low-level interaction honeypot and high-level interaction honeypot.

The **low-interaction honeypot** uses software emulation of network services and operating systems on the host operating system to detect an attacker. The **high-interaction honeypot** functions differently; it permits the attacker to access all services on the given operating system and platform, nothing is restricted.

A **honeynet** is a high-involvement honeypot with the same risks and vulnerabilities that are characteristic for networks of many organizations today. It is "not a single system but a network of multiple systems" [2]. "Honeynets represent the extreme of research honeypots. They are high interaction honeypots, which allow learning a great deal; however they also have the highest level of risk. Their primary value lies in research and gaining information on existing threats. A Honeynet is a network of production systems. Nothing is emulated. Little or no modifications are made to the honeypots. This gives the attacker a full range of systems, applications, and functionality to attack. From this it can be learnt a great deal, not only their tools and tactics, but their methods of communication, group organization, and motives" [3].

According to definitions of the honeynet and virtualization, a **virtual honeynet** can be defined as "a complete honeynet, running on a single computer in virtual environment" [3]. A virtual honeynet can be defined as "a technology that virtually implements many different operating systems in one hardware computer, and hence instead of having a honeynet of different physically separate honeypots, all the honeypots will be virtually set in one machine and still appear to the attacker as different separate machines" [4]. Virtual honeynets combine all the elements of a honeynet into a single physical system. Not only are all of the three requirements of data control, data capture, and data collection met, but also the actual honeypots themselves run on the single system [5].

A successful deployment of a honeynet is a successful deployment of its architecture. There are some core elements of the honeynet architecture [2]:

- **Data control** is the first requirement whose purpose is to control and contain the activity of the attacker.
- **Data capture** monitors and logs all of the attacker's activities within the honeynet.
- **Data collection**—in case when the organization has more than one honeynet, all data has to be captured and stored in one central location.
- **Data analysis** is an ability to analyse the data collected from the honeynet.

Deployment and usage of honeynets may lead to a number of problems and issues. This paper outlines **legal issues** affecting the deployment and usage of the honeynets. Legal analysis is based on the European Union law (the EU law). Paper discusses the European Union regulations, the EU directives and international agreements. National legislation of member states of the EU are based on these legal documents (the EU directives, international agreements) or legal documents are an integral part of national legislation (the EU regulations, international agreements).

Therefore, some native legislation may be slightly different from the concept of the EU law or international law.

There are several contributions of this paper. The first contribution of this paper is the review of the research literature related to legal aspects of honeypots and honeynets. The second contribution of this paper is the legal analysis of the core elements of honeynets from the perspective of the EU law. In the case of data capture and data control we focus on evolution of legal issues of these core elements according to generations of honeynets.

To formalize the scope of our work, we state **three research questions**:

- Which legal issues are related to the core elements of honeynets?
- What are definitions of honeypots and honeynets from the perspective of the EU law?
- What are the legal issues of data capture and data control according to generations of honeynets?
- What are the legal issues of data collection, data analysis and data presentation?

This paper is organized into nine sections. In Sect. 2 paper focuses on the papers related to legal issues of honeypots and honeynets. Section 3 outlines the honeynet generations. Section 4 is introduction to legal issues of honeypots and honeynet. This section outlines the definitions of honeypot and honeynet. Sections 5 and 6 contain legal analysis of core elements—data capture and data control according to honeynet generations. Section 7 focuses on data collection and its legal issues. Section 8 outlines data analysis and data presentation and outlines their legal issues. The last Section contains conclusions and author's suggestions on the future research.

2 Related Research

Since honeypots and honeynets are spreading all over networks, more and more authors are interested in different issues, including deploying issues and legal issues.

Spitzner [6] outlined legal issues of honeypots for the first time. He discusses three main legal issues related to honeypots from the U.S. perspective, namely entrapment, privacy and liability. According to him the discussion of an entrapment is the simplest, because "honeypots are not a form of an entrapment" [6]. The usage of honeypots has the most influence to the privacy issues. The last issue is liability. He considers the possibility of being sued if the honeypots are used to harm others.

Mokube [7] outlined the aspects that shall be considered when and deploying and usage the honeypots and honeynets in the United States. These authors include three legal issues that must be considered: entrapment, privacy and liability. In spite of the fact that the administrators have responsibility to secure their networks, their rights to monitor all activities in networks may have some limits. These limits can be found in laws. These authors suggested that the honeypot may be misused to harm other networks and computers. It is related to the liability of the attackers and administrators.

Salgado [6] outlines some of the potential legal pitfalls of operating a honeypot in the United States. He also states that some legal issues have to be taken into account. He recommended that the laws that restrict rights to monitor and record all activities in computer networks take into consideration. He also outlines liability of administrator in case of abuse the honeypots to harm other servers and networks. Salgado thinks that issue of entrapment is overstated [6].

On the other hand **Scottberg** et al. [8] considers the honeypots as an entrapment. It may be relevant to privacy law. The attackers' files are not protected since attackers cannot have legitimate accounts or privileges. The third legal issue is liability. Similarly to the previous authors, they state that if the honeypot is compromised it can be used to attack other computer networks and systems. In this case, the administrator of the honeypot may be liable in due diligence of assets.

All above mentioned authors discuss legal issues of honeypots and honeynets from U.S. law perspective. In this paper we will focus on legal issues of honeypots and honeynets from the EU law perspective. The first example of EU law perspective is **Dornseif et al**. [9]. They concentrate more on criminal and civil liability in honeypots and provide an overview about the most pressing issues concerning German laws.

In **Sokol** [10] author outlines the legal issues of honeypots according to honeynets' generations from The European law perspective. This paper discusses legal issues of honeynets considering their generations. Paper focuses on legal issues of core elements of honeynets, especially data control, data capture and data collection. Paper also draws attention on the issues pertaining to privacy and liability. The analysis of legal issues is based on the EU law. This paper is based on it.

Paper [11] discusses the liability issues of honeypots and honeynets. It deals with civil and criminal liability. This paper also focuses on cybercrime and liability of attackers. Paper [12] focuses on privacy issues of honeypots and honeynets. The paper discusses legal framework of privacy, legal ground to data processing and data collection. The analysis of legal issues is based on the EU law and is supported by discussions on privacy and related issues.

3 Honeynet's Generations

According to the adopted technologies, methods and activities of honeynet's core elements, the honeynet has evolved through three generations (**architectures**), whose description are outlined in following subsections. The core elements contained in each generation is shown in Table 1.

Table 1 The core elements in honeynets' generations

Generation/core element	Data capture	Data control	Data collection	Data analysis/Data presentation
Generation I	X	X	–	–
Generation II	X	X	X	–
Generation III	X	X	X	X

3.1 Description of the 1st Generation of Honeynets

The 1st generation of honeynet was the first honeynet that was developed in 1999. This honeynet has simple methodology, limited capability, and it is highly effective in detecting the automated attacks. It runs at the third ISO/OSI layer. For this reason, the 1st generation of honeynet was implemented in an isolated network. It consists of two elements, such as data capture and data control. Schema of the 1st generation of honeynet is shown in Fig. 1.

3.2 Description of the 2nd Generation of Honeynets

The 2nd generation of honeynet was designed in 2002 by Honeynet Project. It presents more complex honeynet to deploy and usage. This honeynet runs on the second ISO/OSI layer and it inspects the outbound data and blocks or modifies data. Main aim of the 2nd generation of honeynet is to ensure that compromised honeypots of the honeynet are not used to attack the computers and other devices outside the honeynet [6], and to ensure efficient data capture and to make the

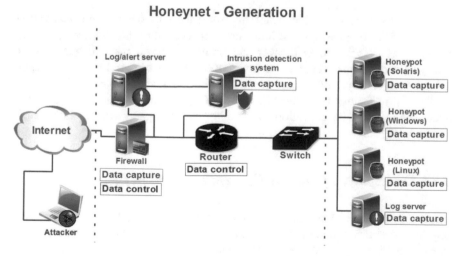

Fig. 1 Schema of the 1st generation of honeynets

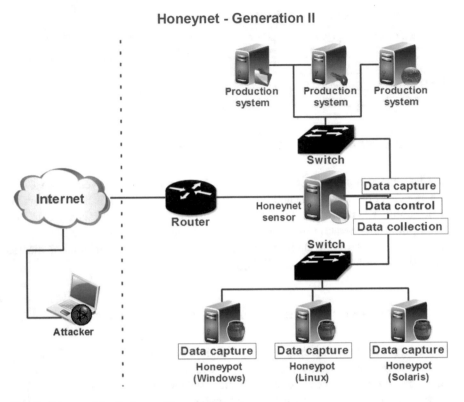

Fig. 2 Schema of the 2nd generation of honeynets

honeynet hard to detect. However, this architecture is different from previous generation of honeynet.

The 2nd generation of honeynets improves the data control and data capture modules of the 1st generation of honeynets. In addition to the previous generation of honeynet IPTables firewall layer and Snort IDS layer, the 2nd generation of honeynet adds another data capture layer where data is captured in the honeypots, using Sebek and Syslogd to capture more information about attackers [3].

The 2nd generation of honeynet consists of three elements, such as data control, data capture and data collection. Schema of the 2nd generation of honeynet is shown in Fig. 2.

3.3 Description of the 3rd Generation of Honeynets

The 3rd generation of honeynet was released at the end of 2004 and it is still used in physical or virtual version [13, 14]. This generation of honeynet shares the same architecture with its predecessor. The 2nd generation of honeynet and the 3rd generation of honeynet differ in deployment and management. The 3rd generation

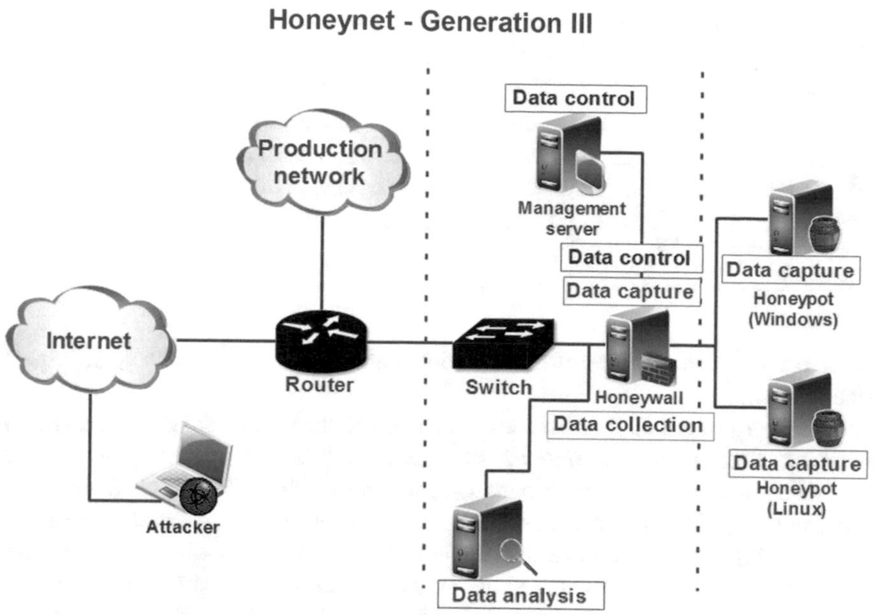

Honeynet - Generation III

Fig. 3 Schema of the 3rd generation of honeynet

of honeynet represent a more refined version of the previous honeynet architecture, unlike the drastic evolution, as it is from the 1st generation of honeynet to the 2nd generation of honeynet. The 3rd generation of honeynet offers a "way of analysis of data from different sources without the person being obliged to go through different data sources manually and to try to determine the relationships" [6].

The 3rd generation of honeynet has brought about new and radical improvements. It contains the core of the 2nd generation of honeynet data control and data capture functionality, but it also has a remote administration of Graphical User Interfaces, data analysis integration, support for the Sebek branch, robust operating system base, automated updating, and much more.

This architecture merges 3 interfaces on the Honeywall. Two interfaces are used as a bridge between the internal Honeypot network and the external network; "whilst the third interface was used for management and configuration tasks" [15]. Scheme of the 3rd generation of honeynet is shown in Fig. 3.

4 Introduction to Legal Issues of Honeypots and Honeynets

There are no legal definitions, according to available information, to what honeypot/honeynet is. Since there is no trace of it even in the EU or U.S. case law, we are locating legal position of honeynets/honeypots by combining technical aspects

Table 2 Legal issues of honeynets according to core elements

Core element/legal issue	Privacy	Liability	Jurisdiction and applicable law	Sharing and publishing	Digital evidence
Data capture	X	–	–	–	–
Data control	–	X	–	–	–
Data collection	X	X	X	–	–
Data analysis and presentation	X	–	–	X	X

(data control, data capture, data collection and data analysis) with legal aspects (the EU law).

According to article 2 of Framework Directive the **electronic communications network**, that means transmission systems and, where applicable, switching or routing equipment and other resources, which permit the conveyance of signals by wire, radio, optical or other electromagnetic means, including satellite networks, fixed (circuit and packet-switched, including Internet) and mobile terrestrial networks, electricity cable systems to the extent that they are used for the purpose of transmitting signals, networks used for radio and television broadcasting and cable television networks, irrespective of the type of information conveyed. Next important definition within the meaning of Article 2 of Framework directive is **electronic communication service**. It is defined as a service normally provided for remuneration which consists wholly or mainly in the conveyance of signals on electronic communications networks, including telecommunication services and transmission services in networks used for broadcasting, but exclude services providing or exercising editorial control over, content transmitted using electronic communication networks and services.

On the basis of these definitions the **honeypot** can be considered a communication service and therefore **honeynet** can be defined as a number of electronic communication services with the electronic communication network.

Design, deployment and usage of each core elements depend on legal issues. Table 2 shows the legal issues (privacy, liability etc.) in relation to the each core element of honeynet. They are discussed in the following sections.

5   Data Capture

The first core element related to honeynet deployment is **data capture**. Primary goals of data capture is monitoring and recording of the conduct of intruder involved in honey net. Use of data capture enables to collect all the information which is contained in data traffic. Note that process of data capture takes place secretly. Data captured from the invader involve all the sending and all receiving information. This provides material for further analysis of the nature and specifics of attacks carried out by the invader. The peculiar challenge lies in capturing as much data as possible, without the process being detected by the attacker [3].

In data capture, the data packet includes record of attacker's behaviour which consists of a set of specific actions. This information will serve as a mean for administrators or researchers for further analysis of methods, strategies, targets or other components of their conduct and contribute to understanding of their behaviour for purpose of designing proper responses. To be more specific, this process involves two functions. The first one is to obtain information on all the changes of network data packets in honeynet. Second function is based in finding and signalling unstandardized behaviour or anomalies and further alerting researchers and administrators [16]. To shed more light on this issue, let us turn our attention to Honeynet Project definition.

5.1 Issue of Privacy

Data capture is closely linked to **issue of privacy**. It is defined as the right to be left alone and to have a private life and also as the right of a person to be free from unwarranted publicity [17, 18]. This right, set out in Article 8 of the European Convention on Human Rights, includes some individual privacy, namely privacy of the home and office, protection of physical integrity and privacy of communication means (telephone calls, chats, e-mails and so on). The legal framework of the EU relevant to honeynets in privacy issue consists of two directives below:

- the EU Directive 2002/58/EC is concerned with the processing of personal data and the protection of privacy in the electronic communication sector, commonly known as the EU Directive on privacy and electronic communications (**e-Privacy Directive**);
- the EU Directive 95/46/EC on the protection of individuals with regard to the processing of personal data and on the free movement of such data, (**the EU Data Protection Directive**).

The EU Directive 2006/24/ES on the retention of data generated or processed in connection with the provision of publicly available electronic communications services or of public communications networks (**The EU Data retention Directive**) focused on the obligations of the providers of publicly available electronic communication services, or of public communication networks. The Court of Justice of the European Union in joined Cases C293/12 and C594/12 annulled this directive on 8th April 2014.

A more detailed analysis of privacy issues in honeypots and honeynets can be found in [12]. In this paper authors focus on legal issues of privacy according to honeynet's generations.

5.2   The 1st Generation of Honeynets

The aim of data capture of the 1st generation of honeynet is to capture as much data as possible, together with that this action is not seized by the attacker. Any system changes to honeypots may be and potentially will be detected by the attackers [5].

All of the network traffic is passing through honeynet's firewall, so logging of the activity is firstly done there (the **first layer**). It logs all incoming, outgoing and related connections from and to honeynet and can notify administrator if applicable. Capturing attackers' keystrokes is not supported. Structure of logs created with firewall is following: packet headers (source IP address and port, destination IP address and port), the date of the attack etc.

Intrusion detection system (IDS, e.g. snort) is used as the **second layer** of data capture. It's job is to notify the administrator of chosen services and to give more specific information [1]. Snort can seize all activity on given network, store logs of it in ASCII and binary format and afterwards send all alerts to a logging server using some sort of script [3]. An IDS has two interfaces, one is connected to the honeynet and the other interface is connected to the internet or some other network for remote administration and data collection purposes.

The **third layer** of data capture is represented by the actual honeypots. Because of modifications in shell (e.g. bash or ComLog) commands can be captured and screenshots can be made. Since acquired logs are sent to different (logging server) the communication "must be as secure and discrete as possible" [18].

Legal issues of data capture are based on two aspects, namely what data are collected and legal grounds of collecting them. There are two categories of these data:

- contents of communications
- information to establish communication

Article 2(a) of e-Privacy Directive states **communication data** are any data transferred or exchanged between a final numbers of attendants by any means of a generally available electronic communication service. Content (communication) data are all possible file contents, such as bodies of emails messages, content from active sessions of attacker (e.g. attackers' keystrokes) etc.

According to the article 2(2a) of this Directive transactional data (**no-content data**) are traffic data, location data and the related data necessary to identify user, such as IP addresses, connection ports, network protocols, account names, urls etc.

Legal grounds to process data is another issue of privacy. E-Privacy of EU Directive states that the relevant legal ground for dealing with the data in honeynets is assuring the security of the computer or network if honeypots in honeynet are production. The case when honeynet consists of research honeypots is different. The administrator of that system can use other possible grounds to legitimise and justify managing policies based on transactional or content data. Previous agreement is necessary in this type of honeynet.

To capture all the data from all possible users and not only from attackers, it necessary to place some kind of notice (e.g. banner) about data collecting. The **banner** can be used as user-consent in case of honeynet, which captures the personal data. User consent to legitimise the processing can be found in article 5(1) of the e-privacy Directive. It has to be "free, specific, informed and indication of wishes" [19]. Our recommendation is to put this text into user-consent (banner): This system is intended to be used by authorized users only. By proceeding further you are giving your consent to record, store and unveil all your activities on this system.

We consider using user-consent in case honeynet placed in production environment, since both types of data (content data and transactional data) are collected. The reason for this is the fact that consent must be obtained from each user participating in a communication [19]. The **most problematic layer** of data capture in the 1st generation of honeynet is **firewall**. All network traffic from both production network and honeynet, must flow through firewall. It is critical place from the perspective of legal issues of data capture. Administrators of honeynet must have consent from all users in production network to capture all network traffic. From this perspective, this generation of honeynet is useless.

5.3   The 2nd and the 3rd Generation of Honeynets

The 2nd and the 3rd honeynets' generation use similar mechanisms for data capture as the 1st one. In spite of this, in the 2nd and the 3rd generation of honeynets there are more sophisticated and more advanced approaches which are more reliable and their detection is more difficult. This is mainly due to modifications of kernel modules, which are able to record the actions without attackers' knowledge.

As in 1st generation of honeynet, to seize data from all (not just attackers), it would be proper to notify users (e.g. using some kind of banner) about their data being collected. Banner (user-consent) mentioned in 1st generation of honeynets is also applicable in these generations of honeynets.

The second part of data capture sensors are honeynet sensor or honeywall. These components seize network traffic data only from the honeynet itself, not from the rest of production network.

6   Data Control

The main purpose of data control is to ensure that the compromised honeypot cannot be used to attack other network. This is the most important function and it always has to be given high priority when implementing a honeynet [3]. Honeynet data control can reduce the risk of being misused by intruders. Once the intruders take over one of the honeynets, the intruder can use the compromised honeypot to attack other hosts, so it is necessary to control the attackers [6].

Some examples of data control approaches include counting the number of outbound connections from within the honeynet. Data control should be configurable by the administrator at any time, including remote access for cases, when a problem arises, when the administrator is not physically in the honeynet area. There should always be automated alerting, when a honeynet is compromised [3].

6.1   Issue of Liability

Data collection is closely linked to issue of **liability**. It is referred to as a legal duty or obligation [16]. Liability is one of the most important words in the law field. It is defined as a legal responsibility for one's acts or omissions. If a person (entity) fails to meet that responsibility, it leaves him open to an action from court such as lawsuit or a court order [17].

Honeynet's administrator liability can be defined as a legal responsibility for acts done by him (e.g. incorrectly setting the honeynet) or omissions (e.g. underestimating the attacker) associated with the design, deployment or usage of honeypots.

These directives applicable to honeynets are stated in the EU legal framework:

- Directive 2002/21/ES on a common regulatory framework for electronic communication networks and services (**Framework Directive**);
- Directive 2002/58/EC of the European Parliament and of the Council of 12 July 2002 concerning the processing of personal data and the protection of privacy in the electronic communications sector (**the EU E-privacy Directive**);
- Directive 98/34/EC establishes a procedure for the provision of information in the area of technical standards and regulations;
- Directive 2000/31/EC on certain legal aspects of information society services, in particular electronic commerce, in the Internal Market ('the EU Directive **on electronic commerce**').

A more detailed analysis of liability issues in honeypots and honeynets can be found in [11]. In this paper authors focus on legal issues of liability according to honeynet generations.

6.2   The 1st Generation of Honeynets

The 1st generation of honeynet has relatively simple approach to data control. It consists of two layers, namely firewall and router, which are also used for data capture. Another requirement for data control is security monitoring. Administrators of honeynet have to be informed about any violation of data control requirements (ingoing connection is allowed). The basic tool for data control is firewall. It allows any ingoing connections, but controls all outgoing connections. If a certain threshold of outgoing connections is reached, the firewall blocks all other attempts. To avoid the single point of failure the honeynet does not depend on a single layer for data control, there is also a router as a second layer of data control.

Administrator of honeynet is responsible for the **adequate network protection**. This responsibility can be found in several provisions of acts. The EU law requires **sufficient degree of security**. First example is Article 13a of the EU directive 2009/140/EC that outlines the security and integrity of networks and services. According to this article, the honeynet administrators take appropriate technical and organisational measures to manage the risks posed to security of networks and services appropriately. Another example of this requirement is Article 4(2) of e-privacy the EU Directive. According to this legal statute, in case of a particular risk of a breach of the security of the network, the honeynet administrator provider must inform production network users of such a risk.

Main problem of this generation of honeynet is approach to block activities of attackers. If a number of connections has reached a certain threshold, the firewall would block all further attempts. For example, after 10 attempts firewall blocks connection from infected honeypot. Data control carried out its activity, but honeypot attacked non-honeypot system and it might cause damage. In this respect, the data control in the 1st generation of honeynet does not represent adequate level of security required by the EU law. This implies that it is not suitable for deployment.

Example Attackers have taken over a system within the honeynet; they can attempt to launch exploits against a non-honeynet system (e.g. web server). In this case, the data control is limited. After several attempted outgoing activity, all further activity (including any exploits) is blocked. In such case the attack is carried out (within outgoing activity). Liability of the honeynet administrator is based on type, amount and time of attack. If these aspects are sufficient for a successful attack, the honeynet administrator is responsible because design of honeynets includes conditions for attack to non-honeynet system. Adequate level of security required by the EU law is not satisfactory in these cases.

6.3   The 2nd and the 3rd Generation of Honeynets

In comparison with 1st generation of honeynets, **data control** in the 2nd generation of honeynets uses a single honeynet sensor that combines functionality of both firewall and intrusion detection system's sensor. Honeynet sensor is almost impossible to detect due to absent TTL decrement of system hops and no MAC device numbers. The data control is passed in second OSO/OSI layer frames from the production network to the honeynet. The data control is placed "between the honeynet and the system, which captures and controls packets" [1]. How the honeynet handles malicious and not authorized activity is the most significant change. It tries to adjust and restrict the attacker's on the honeynet, not just simply block connection.

All core elements are combined into one device, contrary to the 1st generation of honeynets. This makes data control, data capture and data collection done from one place (computer), which makes it simpler to deploy and manage. Consisting of second ISO/OSI layer gateway this resource operates as a bridge [3]. Another advantage is the fact that the device is more hidden from attackers. Therefore they cannot know that their traffic is being controlled and captured. All traffic, ingoing and outgoing, must flow through one device (gateway).

As we have mentioned before, affecting legal issues are significant changes in the 2nd generation of honeynet are data capture and data control connected in single resource [1] and honeynet location within production network. Since honeynet is located in production network, there has to be increased requirements for administrator to protect other sections of this network. Liability of administrator is based on how many security actions does he take.

Example Attacker has taken control over a system within the honeynet, they can attempt to launch an exploit (e.g. web exploit) against a non-honeynet system (server or computer network). The exploit attempt would be identified and then modified so that attack would be ineffective. The data control in this generation of honeynet would disable or modify exploit code and subsequently it allows modified attack to continue. The attacker sees the launched attack (the packets return), but he cannot understand why the exploit has never worked properly. The honeynet is also able to fake response. For example it blocks connections and forges a dropped connection. In this generation of honeynets the liability of honeynet administrator is comparable to the 1st generation of honeynets. Compared to the 1st generation, the design of honeynet is safer and, in some cases, there is no outgoing traffic from honeynet. Still, we incline to opinion that adequate level of security required by the EU law is also not satisfactory in this generation of honeynet. Few connections or uploads can cause damage and thus liability of administrator.

Since the 2nd and 3rd generations share the same architecture, the discussion of the 2nd generation of honeynet applies to the 3rd honeynet generation as well. In addition, we suggest various improvements. This system shall be based on the internal sandbox, internal network and services [11] and a component similar to the e-mail whitelist. According to our opinion, usage of sandbox [20, 21] will enhance security in honeynet. Requests from honeypots will be redirected to the sandbox and internal network and services and subsequently out of honeynet. Furthermore, the system will have a non-detection feature.

7   Data Collection

The third of the core elements is **data collection**. This function is unique in the way that it is not a requirement for standalone honeynet deployments [5]. If organization deploys a honeynet, it does not need to deploy this functionality. Data collection is necessary only when the organization deploys or manages more than one honeypot.

The purpose of this function is to capture and aggregate all the information from multiple honeynets in one place. This is defined as distributed honeynet.

This core element is mentioned for the first time in the 2nd generation of honeynets. It is closely related to distributed honeynets. The most important issue of data collection is to ensure that the information is collected from honeynets in a safe manner [5].

7.1 Geographical Location of Data

With the deployment of the distributed honeynets, there are few legal issues. First legal issue is related to **geographical location of data**. It reflects the fact that different quality of legal protection is associated with different data protection and privacy settings among various countries. This statement is demonstrated by the fact that data, which are sent out of the country, are protected by applicable law of that country even after leaving this country. The nature of data protection is thus claimed to be extraterritorial [22, 23]. Once the EU Member State's data protection law attaches to personal information, there is no clear way to remove the applicability of the law to the data [24]. Applicability of the law is strongly connected to the data.

This implies the fact that personal data which are governed under the EU law do not lose their protection after leaving the EU territory. Citizens' data cannot fly freely to other jurisdictions without guarantees of the continuation of their protection and application of privacy standards which are adequate from perspective of the EU law. The EU law offers protection beyond territory of the EU, even if the data leave the EU.

In addition to previous paragraph, there is to note that from a legal perspective of contract law, flexibility of data location potentially challenges the governing law provision in the contract. If the law specified in the contract (e.g., the contract says that laws of Thailand will govern this agreement) requires a certain treatment of the data, but the law of the jurisdiction where the data resides (e.g., data center in Poland) requires another treatment, there is an inherent conflict that must be resolved. This conflict exists regardless of whether the storage is temporal, and as part of the processing of the data, or long-term storage that might be a service in itself (i.e., infrastructure as a service), or part of a software or platform as a service offering [24].

7.2 Jurisdiction and Applicable Law

Another issue related to distributed honeynet is **issue of jurisdiction**. Jurisdiction is defined as a court's authority to judge acts committed in a certain territory. **Distributed honeynets** can be located in more than just one jurisdiction. Their components may be located in various countries. Distributed honeynet as such may be of multinational nature and it is possible that it will be governed by different legal orders. Subsequent problem related to overlapping jurisdictions means that in cyberspace, rules of several sovereigns merge into new legal space of legal pluralism. This is due to the fact that distributed honeynets are decentralized and it is possible that they will not fall under one jurisdiction. This raises concerns which are associated with differences in legal regulations and multiple legal standards, which have to be applied in different situations, because cyberspace is, according to different circumstances, governed by different legal regimes. We discuss the civil and criminal jurisdiction issues in subsection called Data control. We focus on these issues in more detail in future research.

7.3   Civil Law

Jurisdiction in case of civil liability in honeynet is based on the EU **Regulation No 1215/2012** of the European Parliament and of the Council of 12 December 2012 on jurisdiction and the recognition and enforcement of judgments in civil and commercial matters. This Regulation shall apply in civil and commercial matters whatever the nature of the court or tribunal. It shall not extend, in particular, to revenue customs or administrative matters. Subject to this regulation, persons domiciled in a member state shall, whatever their nationality, be sued in the courts of that member state. Persons who are not nationals of the member state in which they are domiciled shall be governed by the rules of jurisdiction applicable to nationals of that state. If the defendant is not domiciled in a member state, the jurisdiction of the courts of each member state shall be determined by the law of that Member State. According to this regulation a person domiciled in a member state may, in another member state, be sued:

- in matters relating to tort, delict or quasi-delict, in the courts for the place where the harmful event occurred or may occur (Article 7(2) of Regulation No 1215/2012);
- as regards a civil claim for damages or restitution, which is based on an act giving rise to criminal proceedings in the court seised of those proceedings, to the extent that that court has jurisdiction under its own law to entertain civil proceedings (Article 7(3) of Regulation No 1215/2012).

With regard to above mentioned facts, **administrator or attacker may be sued**:

- in the court of the member state if administrator or attacker resides in that member state;
- in the court determined by the law of member state, if administrator or attacker do not reside in any member state;
- in the court for the place where the harmful event occurred or may occur in matters relating to tort, delict or quasi-delict and

- in the court seized of proceedings, to the extent that that court has jurisdiction under its own law to entertain civil proceedings as regards a civil claim for damages or restitution which is based on an act giving rise to criminal proceedings.

Applicable law in case of civil liability in honeynet is based on **Regulation (EC) No 864/2007** on the law applicable to non-contractual obligations (Rome II). This Regulation shall apply, in situations involving a conflict of laws, to non-contractual obligations in civil and commercial matters. It shall not apply, in particular, to revenue, customs or administrative matters or to the liability of the state for acts and omissions in the exercise of state authority (acta iure imperii). Article 4 of this regulation states that unless otherwise provided for in this regulation, the law applicable to a non-contractual obligation arising out of a tort/delict shall be the law of the country in which the damage occurs irrespective of the country in which the event giving rise to the damage occurred and irrespective of the country or countries in which the indirect consequences of that event occur. However, where the person claimed to be liable and the person sustaining damage both have their habitual residence in the same country at the time when the damage occurs, the law of that country shall apply. Article 4 of Rome II establishes the concept of **lex loci damni** as the general rule. Concept ex loci damni denotes the place where the loss is sustained [25].

According to the above, the law applicable to non-contractual obligation of the attacker and administrator arising out of a tort/delict (e.g. attack, omission) shall be the law of the country in which there is no-honeynet system and where damage was caused by attack. Exception is the when the administrator or the attacker and the person sustaining damage both have their habitual residence in the same country at the time when the damage occurs. In this case the law of habitual residence country shall apply.

7.4 Criminal Law

Last aspect is issue of jurisdiction in criminal law. Within the meaning of the EU **Directive 2013/40/EU** of the European Parliament and of the Council of 12 August 2013 on attacks against information systems and replacing Council Framework Decision 2005/222/JHA, member state of the EU has jurisdiction in some cases.

The first case is when the offence has been committed against **honeynet located in member state** of the EU or against **one honeynet within distributed honeynet** that is located in member state of the EU.

The second case is similar to the first case. Member state of the EU has jurisdiction in case that offence has been committed against no-honeynet system located in member state of the EU from honeynet or against one no-honeynet system located in member state of the EU within distributed type of offense and **honeynet**

is part of this offense (e.g. honeynet is part of the DDoS attack). Within these cases, the member state of the EU has also jurisdiction in following cases:

- The attacker commits the offence when **physically presents on its territory**, whether the offence is against honeynet or not, or no-honeynet system is on its territory—attacker commits the offence against honeynet or no-honeynet system from honeynet from territory of member state of the EU.
- The offence is **against honeynet or no-honeynet system on its territory**, when physically present on its territory, whether the offender commits the offence or not, attacker commits the offence against honeynet or no-honeynet system located in the EU.

Other case is when the offence has been committed against honeynet or against no-honeynet system located **outside the EU from honeynet by the national of the member states** of the EU, at least in cases where the act is an offence where it was committed (e.g. in country joined the Council of Europe Convention on Cybercrime).

8 Data Analysis and Data Presentation

Data analysis is a process that involves the analysis and correlation of multiple types of data at multiple layers. The purpose and value of data analysis is being able to extract different types of data and then turn that data into valuable information [3]. In a several papers [26, 27] authors add data analysis to the above-mentioned core elements. Data analysis is an ability of honeynet to analyse the data, which is being collected from it. Data analysis is used for understanding, analysing, and tracking the captured probes, attacks or some other malicious activities [6]. Example of this core function is combination of security devices, such as firewall (e.g. IPtables), intrusion prevention system (e.g. Cisco IPS) and intrusion detection system (e.g. Snort, Suricata), where these security devices can analyse the network traffic in detail, and return the result of analysis in a visible way. **Data presentation** represents a part of honeynet, where analysed data are visible for honeynet administrators. These data represent a main aim of honeynet activities and present results of activities of the entire honeynets.

8.1 Sharing and Publishing Network Traces

To share and to publish network traces is important matter due to multiple reasons. For example common datasets can be helpful when comparing different experimental approaches, current data may not follow current threats or traffic aspects and simulated data are inadequate for certain reasons [28]. Also it is necessary to mention the **anonymization issue**. The collected data which are going to be presented must be anonymized. This is an active area of study in the security community, as witnessed by the continuous development of anonymization methods and releases of network data that are enabled by them [29]. E-Privacy Directive discusses the anonymous data. According to Article 6 of this directive, the data related to users of a public network or electronic communication service handled and stored by the provider (honeynets administrator in this case) have to be deleted or anonymized when they are no longer required. Publishing of results can also damage an institution's reputation by revealing infrastructure details (computers and networks) the organization wants to keep classified [30].

8.2 Digital Evidence

Since honeynet is a network forensics tool, results of activities of honeynets represent digital evidence. An **evidence** can be defined as information or signs, indicating, whether a belief or proposition is true (valid). Evidence is also information used to establish facts in a legal investigation or admissible as testimony in a law court [31]. Any documentation, which satisfies the requirements of evidence in a proceeding, but which exists in electronic digital form, is defined as **digital evidence** [32]. Digital evidence of an incident is digital data containing reliable information supporting or refuting a hypothesis about the incident being investigated. An object is evidence because it has played a role in an event that supports or refutes an investigation hypothesis [33]. **Network-based evidence** is digital evidence produced as a result of communications over a network. An example of this kind of evidence includes logs, network traffic, browser activity etc. [32].

Above mentioned the EU Directive 2013/40/the EU on attacks against information systems do not focus on digital evidence investigation, but **Convention on Cybercrime** outlines the fundamental framework for digital evidence investigation. Since the most states are signatories of the Convention on Cybercrime, we focus on this document. Section 2 of this Convention states some aspects of procedural criminal law in relation to the criminal investigation and proceeding of cybercrime.

Under Article 16 of the Convention on Cybercrime competent national authorities (e.g. law enforcement) are authorized to obtain the expeditious preservation of specified computer data, including traffic data, that has been stored by means of a honeynet, in particular where there are grounds to believe that the computer data is particularly vulnerable to loss or modification. If competent national authority instructs a person to preserve specified stored computer data in the person's possession, the person is required to preserve and maintain the integrity of that computer data for a period of time as long as necessary, up to a maximum of 90 days, to enable the competent authorities to seek its disclosure. Honeynet administrator is also to keep confidential the undertaking of such procedures for the period of time provided for by its national law. National competent authorities are entitled to search or similarly access:

- a honeynet and data stored therein and
- a computer-data storage medium (e.g. log server within honeynet) in which computer data may be stored in its territory.

Under Article 19 of Convention of Cybercrime national competent authorities are entitled to:

- seize or similarly secure a honeynet or part of it, e.g. its disk, memory and other storage medium;
- make and retain a copy of those computer data;
- maintain an integrity of the relevant stored computer data and
- render inaccessible or remove those computer data in the accessed honeynet.

Within real-time collection of computer data and its existing technical capability, the honeynet administrator is obliged to:

- to collect or record through the application of technical means on the territory of that states and
- to co-operate and assist the competent authorities in the collection or recording of traffic data in real-time, associated with specified communications in its territory transmitted by means of a computer system.

The above mentioned text is related to collection of digital evidence by national competent authorities. The use of this evidence in criminal proceedings, or in proceedings in court is in the European continental legal systems based on the principle of free introduction and free evaluation of evidence. Other principle is the fact that all means of evidence, irrespectively of the form they assume, can be admitted in legal proceedings [34]. From this perspective, all digital evidence obtained in honeynet can be used as evidence in criminal proceedings, or in proceedings in court and his evolution is based on above mentioned principles.

9 Conclusion

Honeynets are interesting interdisciplinary research issues in technological and legal sense. In this paper authors focus on legal issues of honeynets from the EU law perspective. Honeynet consists of four core elements, namely data capture, data control, data collection and data analysis. Deployment and usage of honeynet is related to several legal issues, especially privacy, liability, jurisdiction, applicable law, sharing and publishing traces and digital evidence.

As we have mentioned before, there is no legal definition of honeypot and honeynet. From the perspective of the EU law the honeypot can be defined as a communication service and honeynet is as a number of electronic communication services with the electronic communication network.

Data capture is especially related to privacy issues. In the 1st generation of honeynets we consider several privacy issues, especially categories of data and legal grounds to process data. Data capture in the 2nd and the 3rd generation of honeynets is very similar, but the amount and type of data are spread. On the other hand, data control is especially related to liability issues. According to several provisions of EU acts administrators of honeynet are responsible for an adequate network protection. The 1st generation of honeynet has simple architecture based on connection blocking after the exceeding the limits. This honeynet does not represent adequate level of security required by the EU law. On the contrary, the 2nd generation of honeynet increases requirements for administrator to protect production networks. The liability of administrator is based on how many security actions does he take. Lastly, the 3rd generation of honeynet shall be improving by usage of sandbox, internal network and services.

Data collection is closely linked to the geographical location of honeynet and distributed honeynets. The legal issue related to distributed honeynet is jurisdiction and applicable law. In contrast, data analysis and data presentation is linked to the sharing and publishing results of analysis and digital evidence. Before presenting the collected data it is necessary to anonymize them.

The future work will focus primarily on related problems, mainly on the research of other types of honeypots, especially honeytokens. A challenge for future research is to study jurisdiction and applicable law in distributed honeynets based on cloud technologies.

Acknowledgments We would like to thank our colleagues from the Czech chapter of The Honeynet Project for their comments and valuable input. This paper is funded by the Slovak Grant Agency for Science (VEGA) grant under contract No. 1/0142/15, VVGS project under contract No. VVGS-PF-2015-472 and Slovak APVV project under contract No. APVV-14-0598.

References

1. Pouget, F., Dacier, M., Debar, H.: White paper: honeypot, honeynet, honeytoken: terminological issues. Rapp. Tech. EURECOM. 1–26 (2003)
2. Spitzner, L.: Honeypots: Catching the insider threat. In: Computer Security Applications Conference, 2003, pp. 170–179. IEEE (2003)
3. The Honeynet project: Know Your Enemy: Learning about Security Threats, 2nd edn. Addison Wesley (2004)
4. Mairh, A., Barik, D., Verma, K., Jena, D.: Honeypot in network security: a survey. In: Proceedings of the 2011 International Conference on Communication, Computing & Security. pp. 600–605. ACM, ODISHA, India (2011)
5. Joshi, R.C., Sardana, A.: Honeypots: A New Paradigm to Information Security. Science Publishers, USA (2011)
6. Spitzner, L.: The honeynet project: trapping the hackers. IEEE Secur. Priv. Mag. **1**, 15–23 (2003)
7. Mokube, I.: Honeynets—concepts, approaches and challenges. In: Proceedings of the 45th Annual Southeast Regional Conference on—ACM-SE, vol. 45, pp. 321–326 (2007)
8. Scottberg, B., Yurcik, W., Doss, D.: Internet honeypots: Protection or entrapment? In: IEEE 2002 International Symposium Technology and Society (ISTAS'02), pp. 387–391 (2002)

9. Dornseif, M., Gärtner, F.C., Holz, T.: Vulnerability assessment using honeypots. Praxis der Informationsverarbeitung und Kommunikation **27**, 195–201 (2004)
10. Sokol, P.: Legal issues of honeynet's generations. Electronics, Computers and Artificial Intelligence (ECAI), 6th International Conference on 2014, pp. 63–69 (2014)
11. Sokol, P., Andrejko, M.: Deploying Honeypots and Honeynets: Issues of Liability. Computer Networks. pp. 92–101. Springer (2015)
12. Sokol, P. Husák, M., Lipták, F.: Deploying Honeypots and Honeynets: Issue of Privacy. Availability, Reliability and Security (ARES) (2015)
13. Kumar, S., Singh, P., Sehgal, R., Bhatia, J.S.: Distributed honeynet system using gen III virtual honeynet. Int. J. Comput. Theory Eng. **4**, 537–541 (2012)
14. Abbasi, F., Harris, R.: Experiences with a generation III virtual honeynet. In: Telecommunication Networks and Applications Conference, ATNAC 2009. pp. 1–6. IEEE (2009)
15. Misra, R., Renu, D.: Cyber crime investigation and network forensic system using honeypot. Int. J. Latest Trends Eng. Technol. 34–40 (2012)
16. Law, J.: A Dictionary Of Law. Oxford University Press (2015)
17. Black, H.C., Garner, B.A.: Black's law dictionary. West Publishing Company (1999)
18. Bishop, M.A.: The Art and Science of Computer Security. Addison-Wesley Longman Publishing Co. Inc., Boston (2002)
19. Opinion of the European Data Protection Supervisor on net neutrality, traffic management and the protection of privacy and personal data (2012/C 34/01)
20. Willems, C., Holz, T., Freiling, F.: Toward automated dynamic malware analysis using cwsandbox. IEEE Secur. Priv. **5**, 32–39 (2007)
21. Cuckoo Sandbox project. http://www.cuckoosandbox.org (2015)
22. Ustaran, E.: The Scope of Application of EU Data Protection Law and Its Extraterritorial Reach. Beyond Data Protection. pp. 135–156. Springer, Berlin, Heidelberg (2013)
23. Svantesson, D.J.B.: A "layered approach" to the extraterritoriality of data privacy laws. Int. Data Priv. Law **3**, 278–286 (2013)
24. Bowen J.A.: Legal issues in cloud computing. In: Buyya, R., Broberg, J., Goscinski, A. (eds) Cloud Computing: Principles and Paradigms. pp. 593–613. Wiley, Hoboken (2011)
25. Fawcett, J., Carruthers, J.M., North, P.: Private International Law. Oxford University Press (2008)
26. Balas, E., Viecco, C.: Towards a third generation data capture architecture for honeynets. In: IAW'05. Proceedings from the Sixth Annual IEEE SMC. pp. 21–28 (2005)
27. Shi-wei, Y., Xiu-shuang, M., Wei-dong, W.: Core functions analysis and example deployment of virtual honeynet. Comput. Sci. **3**, 21 (2012)
28. Pang, R., Allman, M., Paxson, V., Lee, J.: The devil and packet trace anonymization. SIGCOMM Comput. Commun. Rev. **36**, 29–38 (2006)
29. Coull, S.E., Collins, M.P., Wright, C.V., Monrose, F., Reiter, M.K., et al.: On web browsing privacy in anonymized NetFlows. USENIX Security (2007)
30. Burstein, A.: Conducting cybersecurity research legally and ethically. In: LEET'08 Proceedings of the 1st Usenix Workshop on Large-Scale Exploits and Emergent Threats. pp. 1–8. San Francisco (2008)
31. Oxford Dictionaries Online—English Dictionary and Language Reference. Online: http://oxforddictionaries.com (2015)
32. Davidoff, S., Ham, J.: Network Forensics: Tracking Hackers Through Cyberspace. Prentice hall (2012)
33. Carrier, B.D., Spafford, E.H.: Defining event reconstruction of digital crime scenes. J. Forensic Sci. **49** (2004)
34. Karyda, M., Mitrou, L.: Internet forensics: legal and technical issues. Work. Digital Forensics Incident Anal. Int. **0**, 3–12 (2007) (IEEE)

A Risk Screening System by Network Diagram Recognition for Information Security Audit

Masaki Samejima and Naoki Satoh

Abstract We address risk screening support for the information security audit. In audit planning, the auditor has limited time to collect information and discuss risks with the company to be audited. The efficient way of identifying risk correctly within the limited time to improve the quality of the information security audit. Focusing on a network diagram that is commonly used for risk identification, we propose a risk screening system by network diagram recognition. The proposed system captures a picture of the network diagram by a camera, and stores the picture in volatile memory. The content of the network diagram on the picture is recognized by technologies of computer vision and saved as a XML file. Applying risk identification rules given by the auditor to the recognized network diagram, the proposed system identifies risks on the network diagram.

Keywords Risk screening · Information security audit · Network diagram recognition · Computer vision · Risk assessment

1 Introduction

Due to the increasing security incidents [1], companies try to improve their information security management to prevent the incidents. According to guidelines of information security management system, such as ISO/IEC 270001 [2], companies should plan the policy of information security, do risk assessment and

M. Samejima (✉)
Graduate School of Information Science and Technology, Osaka University,
2-1 Yamadaoka, Suita-shi, Osaka 565-0871, Japan
e-mail: samejima@ist.osaka-u.ac.jp

N. Satoh
Wakayama University, Sakaedani 930, Wakayama-shi, Wakayama 640-8510, Japan
e-mail: nsatoh@center.wakayama-u.ac.jp

© Springer International Publishing Switzerland 2016
E. Pricop and G. Stamatescu (eds.), *Recent Advances in Systems Safety and Security*,
Studies in Systems, Decision and Control 62, DOI 10.1007/978-3-319-32525-5_11

treatment, check whether the risk assessment and treatment are good, and act for improvement based on result of the check. Nowadays, many risks are hidden in large-scale and complex information systems, and targeted by a malicious attacker. Because it is difficult for only companies to find the risks [3], the companies often ask auditors that are third party organizations to check the risk assessment and risk treatment [4].

When an auditor is asked to check by a company, the auditor starts to collect information with the company for information security audit. Interview and document review are often used for collecting the information, such as a security policy, a installed system [5]. Based on the collected information, the auditor identifies risks that are related to information security. For example, if the security policy does not indicate that users should change their passwords for the system frequently, there is a risk that their user accounts may be taken over. After the risk assessment, the auditor submits an audit report and proposes measurements to mitigate the risk. Through discussions between the company and the auditor, the company determines which measurements are to be implemented.

In the above process for information security audit, missing the risks is a critical problem. In order to avoid the problem, the auditor has to collect sufficient information for the risk assessment, but the time for collecting information is limited and the access to the documents is restricted, which makes it difficult to collect sufficient information. The auditor often uses templates and checklists for efficient collecting information. With the support of such tools [5, 6], the auditor collects information several times because necessary information is found through iterative interviews and document reviews. In the iterative interviews and document reviews, the auditor first grasps the perspective of the information system that is installed in the company by a computer network diagram, and next identifies components that have risks in the system. After that, the auditor focuses on and collects more information on the components. However, the auditor sometimes fails to identify the components that have risks.

In this paper, we propose a risk screening system for information security audit based on a network diagram. Due to the security reason, the network diagram is provided as a print media rather than a electronic media. In order to handle the network diagram by information processing, the proposed system transforms the image data of the network diagram into graph data, which is called network diagram recognition. By the auditor's making If-Then rules for screening risks on the graph data, the proposed system can identify the components that have the risks based on the rules.

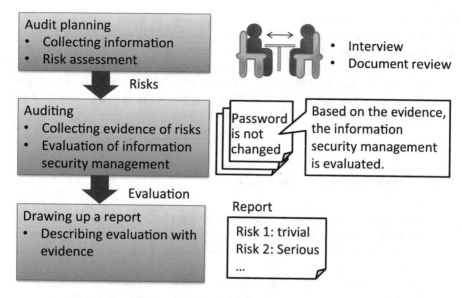

Fig. 1 Outline of audit process

2 Information Security Audit Process

2.1 *Outline of Audit Process*

Figure 1 shows the outline process of information security audit, which consists of audit planning, auditing, and drawing up a report. Each step is described in the following:

Audit planning The goal of audit planning is to determine the purpose of the audit, the scope of the audit, and so on. In order to determine them, the auditor first collects information that is related to audit planning through interviews and document reviews. For example, if an organization to be audited have critical web-based services, the purpose of the audit is to protect the services from malicious attacks and the scope of the audit is the web-based systems including who administer the systems. The auditor interviews administrators and reviews a diagram of the network of the web-based systems. Next the auditor assesses risks based on the collected information. If some risks are identified, the auditor collects more information on the risks; if some vulnerable components are found in the system, the auditor asks the administrators the detail of the components. To collect more information can lead to reveal another risks.

Auditing In the auditing, the auditor collects evidence of the risks that are assessed in planning by interviewing more people, reviewing detailed documents, and using risk detection tools. For the example of vulnerable components in the above, the auditor checks parameter values of the components based on checklists and carries out harmless attacks to reveal the vulnerability. Through this process, the information security management in the company is evaluated.

Drawing up a report Finally, the auditor draws up a report on the risks with evidence. Because the report is used for improving the information security management, the content includes what is necessary for improving it. The report includes not only findings in the audit but also information from the outside of the company, such as a trend of information security.

2.2   Problem in Audit Planning

If the auditor fails to identify a critical risk in the audit planning, the risk is not considered in later steps, which may cause that the company to be audited will be damaged due to the risk. In this paper, we address supporting the risk identification in the audit planning.

In order to avoid the problem of missing risks, the auditor generally uses a checklist to identify risks. The auditor sees an item of a risk in the checklist and finds documents or persons that are related to the item. Through reviewing documents or listening to the persons, the auditor can judge whether the risk in the checklist is in the company or not.

Because the auditor want to use the checklist widely for various kinds of situations of companies, risks in checklists are described abstractly. On the other hand, information from persons and documents in companies are specific to the companies. Therefore, the auditor needs to interpret abstract meaning of the information from persons and documents, and compare the content of the information to the checklist.

In case of complex and uncommon information systems, it is difficult to interpret the information, which makes the auditor fail to identify risks correctly. In particular, a network diagram is useful to grasp an outline of a system installed in the company, but the auditor sometimes fails to identify risks in the network diagram. Figure 2 shows an example of risk identification on a network diagram.

As shown in Fig. 2, a network diagram often includes nodes that indicate servers, firewalls, routers, PCs and so on, and links between nodes. A label for each node is written near the node, but is often omitted when the label is duplicated. In addition, the network diagram sometimes includes information of zones and access control, such as DMZ in Fig. 2. Then the auditor can identify some risks based on the network diagram with risk identification tools, such as a checklist.

However, additional information, such as information of zones, is not always on the network diagram. The auditor has to judge whether additional information is necessary or not for identifying risks. Because the auditor can not spent much time for the audit planning, the auditor fails to identify some risks on the network diagram without collecting additional information. In order to avoid the fails, we propose a risk screening system based on the network diagram.

Fig. 2 Risk identification on
a network diagram

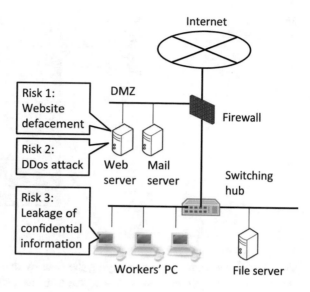

3 Risk Screening System by Network Diagram Recognition

3.1 Requirements

Figure 3 shows requirements for the risk screening system. Because network diagrams are confidential information in general, the network diagrams are provided as a print media in a restricted area of the company to be audited. The auditor cannot take the network diagrams outside the company and store data of network diagrams in the auditor's storage device. Therefore the risk screening system has to work without saving the data in a storage device.

In addition, the time period when the auditor can see the network diagram is limited by the company to be audited. Time-consuming tasks for capturing data of

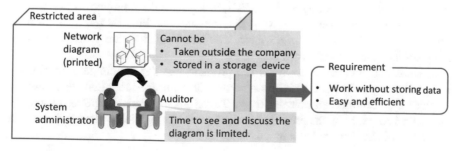

Fig. 3 Requirements for risk screening system

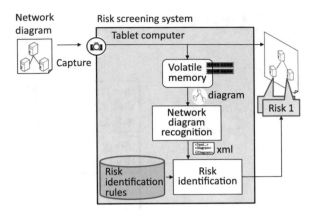

Fig. 4 Outline of the risk screening system

the network diagrams, such as copying the data by manual, are not appropriate for the system. More efficient methods to capture the data is required to realize the system.

3.2 System Design

For efficient risk screening without storing data in a storage device, the proposed system captures a picture of a network diagram in volatile memory, recognizes the content of the network diagram from the picture, and identify risks from the network diagram. Figure 4 shows the outline of the proposed system.

The proposed system is implemented on a tablet computer equipped with a camera. When a picture of a network diagram is captured by the camera, the picture is stored in volatile memory and shown to the auditor. Stored pictures in volatile memory are deleted by shutting down the risk screening system after audit planning in a restricted area. From the picture in volatile memory, the information of network diagram is extracted and transformed to a XML (Extensible Markup Language) file that defines nodes denoting servers, PCs, etc. and links denoting network lines between nodes. Risks on the network diagram are identified by risk identification rules that the auditor makes in advance. The identified risks are superimposed on the picture of the network diagram.

Because our purpose is to avoid the problem that the auditor fails to identify risks, all the possible risks on the network diagram are identified regardless of whether the identified risks exist truly. Therefore, network diagrams should be recognized correctly by the proposed system, and risks should be identified as much recall rate of the identification as possible.

```
<?xml version="1.0">
<diagram>
   <node id = "1", type = "Firewall">
      <linknode>2</linknode>
      <linknode>3</linknode>
   </node>
   <node id = "2", type = "Web server">
      <linknode>1</linknode>
      <linknode>3</linknode>
   </node>
   <node id = "3", type = "DB server">
      <linknode>1</linknode>
      <linknode>2</linknode>
   </node>
</diagram>
```

Fig. 5 An example of a XML file

3.3 Network Diagram Recognition

The network diagram recognition is to transform a picture of a network diagram to a XML file that has the following elements. An example of a XML file is shown in Fig. 5.

diagram is a root element of XML file and consists of nodes in a network diagram. In the example in Fig. 5, there are three nodes in a network diagram.
node includes information of a unique id, type of nodes, and which nodes to be linked. The first node in Fig. 5 is assigned to id 1, its type is a Firewall, and linked to nodes whose ids are 2 and 3.

The problem of recognizing a picture of a network diagram is to identify nodes, their types, and links between nodes. Because the auditor empirically knows which types of nodes should be considered for risk identification, the auditor can give candidates of types in the proposed system. Some of node types can be selected from candidates. The process of recognizing a network diagram is shown in Fig. 6 and described in the following:

1. Text information extraction as node types
   Texts on a network diagram often indicates what symbols indicate. As shown in an example of Fig. 6, a symbol near a word "Firewall" indicates that the symbol is a node and its node type is Firewall. In order to identify node types, these text information is extracted by techniques of computer vision [7].
2. Type correction
   Such as a word "DMZ" in Fig. 6, there are words other than node types on the network diagram. Furthermore, different words are used for the same node type. In order to remove the words other than node types and unify the words for one

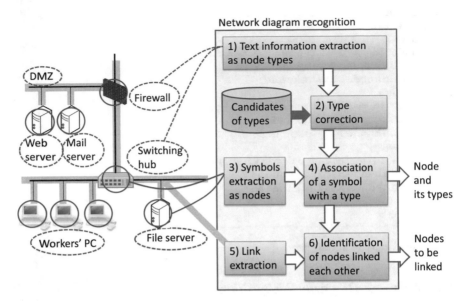

Fig. 6 Network diagram recognition

node type, node types that are extracted from the network diagram are corrected by candidates of node types.

3. Symbols extraction as nodes

The symbols denoting nodes have to be extracted to recognize how many nodes exist and which nodes link to a node. Typically, the nodes are pictures of servers, PCs, routers, and so on. These pictures consist of more edges and colors than any other things on the network diagram. Pattern recognition with these features is considered to be effective for symbols extraction. In addition, pictures that are often used for symbols are similar to each other. If we can specify correct pairs of a picture of a symbol and its node type, machine learning by using the pairs as supervised data is also useful.

4. Association of a symbol with a type

By associating a symbol with a type that have been extracted, it is possible to know nodes and their types and to contribute to identifying links between nodes. The symbol and the type that are associated each other are drawn close to each other on the network diagram. Therefore, the proposed system associates a node and a type that are close to each other. However, duplicated text information that are node types are often omitted, which causes wrong association. Even if duplicated text information is removed, the symbols that are corresponding to the removed text information are still duplicated as shown in an example of PCs in Fig. 6. The proposed system associates the similar symbols with the type.

5. Link extraction

Because links are drawn as lines between nodes, the first step of the line extraction is edge detection by the gradient of the picture [8]. Edges are extracted anywhere on the picture and have to be selected as links. The links tend to be the lines that do not intersect each other and be drawn like branches of a tree. The edges are selected by these features as links.

6. Identification of nodes linked each other

By tracing a line from one endpoint node to another, it is possible to identify the nodes that are linked each other.

3.4   Rule-Based Risk Identification

The auditor identifies risks based on connection patterns of nodes, e.g. if a web server connects to the internet, there is a risk that the content in the web server is defaced. Therefore, rules for the risk identification can be expressed as IF-THEN rules. Let $\mathcal{G}_i$ denote a node set of a connection pattern for identifying risk i. The first and last elements in $\mathcal{G}_i$ are endpoints in connection patterns, and the elements between the first and the last elements are relay nodes. For example, $\mathcal{G}_i = \{\text{Web server}, *, \text{Internet}\}$ indicates any connections from a Web server to the Internet. The symbol * indicates any nodes. The IF-THEN rule is described as follows:

IF The network diagram includes $\mathcal{G}_i$.
THEN There is risk i in the network diagram.

Risks are identified by applying all the risk identification rules to the network diagram. Identified risks are superimposed on the network diagram.

4   Conclusion

In this paper, we proposed a risk screening system by network diagram recognition for information security audit. Because the auditor does not have sufficient time to collect information in audit planning, an efficient way of screening risks is necessary for efficiently collecting information that is related to the risks. Because the auditor identify risks based on the network diagram that is provided as print media for the audit planning, the proposed system recognizes the content of the network diagram through a camera and identify risks by a rule-based method. The proposed system applies techniques of computer vision to recognize the texts, symbols and lines on the network diagram. The texts and the symbols are associated each other, and associated ones are regarded as nodes. By tracing lines between nodes, it is possible to recognize links between nodes. The rule-based risk identification method uses IF-THEN rules to identify risks when the network diagram includes connection patterns of nodes.

As future works, we will implement the proposed system on a tablet computer, and evaluate the effectiveness of the proposed system. The effectiveness is measured by three criteria: accuracy of identifying risks, processing time, and visibility of risks. The accuracy of identifying risks depends on the accuracy of network diagram recognition, which also should be improved. Processing time is taken for network diagram recognition and searching connection patterns in the network diagram. As a result of identifying many risks on the network diagram, it may become difficult to know critical risks. Another thing to be required is to guarantee that the confidential information is not leaked outside the company to be audited. Instead of the auditor's table computer, the company to be audited can install the risk screening software into the company's tablet computer, which reduces the risk that the auditor leaks the confidential information by mistake. The risk screening system should be flexibly adjusted to the requirement of the company to be audited.

References

1. PwC: Managing cyber risks in an interconnected world: key findings from the global state of information security survey (2015)
2. Humphreys, E.: Implementing the ISO/IEC 27001 Information Security Management System Standard. Artech House Inc, Norwood, MA, USA (2007)
3. Huang, S.M., Hung, W.H., Yen, D.C., Chang, I.C., Jiang, D.: Building the evaluation model of the it general control for CPAs under enterprise risk management. Decis. Support Syst. **50**(4), 692–701 (2011)
4. Ohki, E., Harada, Y., Kawaguchi, S., Shiozaki, T., Kagaya, T.: Information security governance framework. In: Proceedings of the First ACM Workshop on Information Security Governance, pp. 1–6 (2009)
5. Stoneburner, G., Goguen, A., Feringa, A.: Risk Management Guide for Information Technology Systems. National Institute of Standards and Technology (2002)
6. Tryfonas, T., Kearney, B.: Standardising business application security assessments with pattern-driven audit automations. Comput. Stand. Interfaces **30**(4), 262–270 (2008)
7. Jung, K., Kim, K.I., Jain, A.K.: Text information extraction in images and video: a survey. Patt. Recogn. **37**(5), 977–997 (2004)
8. Nixon, M.: Feature Extraction & Image Processing for Computer Vision. Elsevier (2012)

Risk Analysis and Optimal Model for Efficiency of Reengineering the Independent Power Grids

Adela Vintea, Paul Schiopu and Octavian Mihai Ghita

Abstract The reengineering of power grid networks is a necessary step that allows us to pass to the next generation of Distributed Power Networks. The frequent use of alternative green energies demand for better equipped networks including elements of smart grids, newer sensors, data transmission capabilities and extensive use of optical fiber. All these efforts imply higher costs, flexibility and scalability to be achieved. This paper present a mathematical model to analyze the efficiency of reengineering taking into account not only the costs, but also associated services and human resources involved. Also, involved risks are analyzed and methods for reducing these risks are presented.

Keywords Reengineering · Optical fiber · Mathematical model · Power grids

A. Vintea (✉) · P. Schiopu
Faculty of Electronics, Telecommunications and Information Technology,
University Politehnica of Bucharest, Bucharest, Romania
e-mail: adela.vintea@ieee.org

P. Schiopu
e-mail: paul.schiopu@yahoo.com

A. Vintea · P. Schiopu
Department of Electronic Technology and Reliability, University Politehnica
of Bucharest, 1-3 Iuliu Maniu Blvd., 061071 Bucharest, Romania

O.M. Ghita
Faculty of Electrical Engineering, University Politehnica of Bucharest,
Bucharest, Romania
e-mail: octavian.ghita@upb.ro

O.M. Ghita
Department of Electrical Measurements, University Politehnica of Bucharest,
313 Splaiul Independentei, 060042 Bucharest, Romania

© Springer International Publishing Switzerland 2016
E. Pricop and G. Stamatescu (eds.), *Recent Advances in Systems Safety and Security*,
Studies in Systems, Decision and Control 62, DOI 10.1007/978-3-319-32525-5_12

1 Introduction

The main difference between today classic networks and future power grids consist in the higher amount of equipment that need to be use, to prepare the network for expected changes. Generalized use of renewable energies imply also that flexible and intelligent networks need to be installed and the logical step is changing or re-engineering the current networks. For example, a wind generator that transfer power to the network should be accustomed by smart sensors for wind speed estimation, more powerful and flexible inverters, robust power transformers and an optical fiber network to allow data traffic. Beside this, human decisions must be taken, knowing also additional information about customers, climate conditions and several other factors [1–3].

It is very difficult to let an automatic hierarchical chain to control such networks, because there are many parameters that cannot be estimated correctly. In such conditions, the total effort to reengineer these networks becomes a burden without having a tool to estimate the efficiency of the process (see Fig. 1).

The efficiency estimation can be performed by a better management of the Independent Power grids (IPG). There is a need for establishing a new solution to draw up the decision through the Command-Control-Communications system for Informatizated management (C^3I) through an Informational Flux of Decision (IFD) (Fig. 2) .

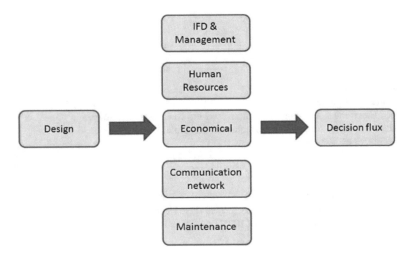

Fig. 1 Decisional model for independent power grid (IPG)

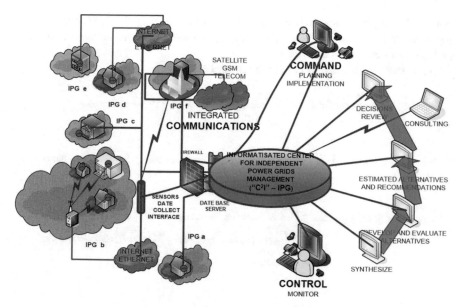

Fig. 2 C³I—IPG system overview

2 Mathematical model

This paper propose a mathematical model to calculate the cost modifications of reengineering process through all steps involved into:

- Design costs of the system;
- Management & IFD costs;
- Maintenance costs;
- Human resources costs;
- Economical costs;
- Decision and communication costs.

The mathematical model for tracking efficiency of C³I—IPG system is as follows [3, 4]:

$$
\begin{aligned}
C_{C^3I-IPG}^{TOT} = {} & C_{\text{design } C^3I \text{ flexibility}}^{TOT} + C_{\text{management \& } IFD}^{TOT} + C_{\text{menten } C^3I \text{ securization}}^{TOT} \\
& + C_{\text{human resource}}^{TOT} + C_{\text{economic financial}}^{TOT} + C_{\text{decision \& } communication}^{TOT}
\end{aligned}
\tag{1}
$$

and

$$
C^{TOT} = \text{total expenses}
$$

where C_i—expenses for each of the functionalities: *design & C^3I-flexibility*; *management & IFD*; *maintenance C^3I-security*; *human resource*; *economic-financial* and *decision & communications*.

Each term from relation (1) can be described as:

$$C_{design\ C^2I\ flexibility}^{TOT} = C_{technical\ scientific\ research}^{TOT} + C_{forecast\ prognosis}^{TOT} + C_{risc\ catastrophe\ chaos}^{TOT} \tag{2}$$

$$C_{management\ \&\ IFD}^{TOT} = C_{control\ \&\ communications\ Master\ \&\ Slave\ system}^{TOT} + C_{reengineering\ \&\ IFD}^{TOT} + C_{marketing\ promotion\ advertisement}^{TOT} \tag{3}$$

$$C_{mentenance\ C^2I\ \&\ \sec urization}^{TOT} = C_{mentenance\ Master\ \&\ Slave\ system}^{TOT} + C_{sensors\ equipments\ repairement}^{TOT} + C_{optimal\ regime\ for\ functioning}^{TOT} + C_{Total\ Qualitiy\ management}^{TOT} + C_{INFOSEC}^{TOT} \tag{4}$$

$$C_{human\ resource}^{TOT} = C_{training\ improvement}^{TOT} + C_{ergonomy\ empaty}^{TOT} + C_{level\ of\ payment}^{TOT} + C_{procedures\ promotion\ standardisatoin}^{TOT} \tag{5}$$

$$C_{economic\ financial}^{TOT} = C_{accounting}^{TOT} + C_{financial\ evidence}^{TOT} + C_{optimal\ regime\ for\ functioning}^{TOT} + C_{level\ of\ fraud}^{TOT} \tag{6}$$

$$C_{decision\ \&\ communication}^{TOT} = C_{early\ warning}^{TOT} + C_{decision\ support}^{TOT} + C_{synthesize\ consulting\ review}^{TOT} \tag{7}$$

Relations (2)–(7) represents the mathematical expression for each term in Relation (1). It is recommended to have a preliminary estimation of each term before starting any numerical calculation of (1). All these considerations should be taken into account in order to understand how logistics monitoring of the C^3I—IPG systems contributes to ensure the success of any reengineering activity. Relation (1) can be improved using weighting coefficients that focuses on one or another step in the process. This can be done establishing some monitoring priorities for our designations.

3 Monitoring Priorities

Because the mathematical model use many terms, each with his own specifications it is very difficult to correctly estimate their values. A monitoring strategy is required in order to obtain some valuable data. This can be achieved using a

Table 1 C³I—IPG
characteristics and services

Basics characteristics		Associated services
Dispersion	a	Decentralization
	b	Flexibility
	c	Independence
Invulnerability	a	Indeterminacy SE
	b	Info security
	c	Survivability
Mobility	a	Modularity
	b	Redundancy
	c	Self-reparability
	d	Good tech design
	e	Homogeneity
Responsiveness	a	Adaptability
	b	Data transformation
	c	Connectivity
	d	Decision support
	e	Direction/monitoring
	f	Knowledge maintenance
	g	Relevancy
Opportunity	a	Early warning
	b	Execution time
	c	Reliability

prioritization of C³I—IPG. The C³I—IPG priorities means to establish, define and determine the main requirements of the structure [4, 5] (Table 1).

Calculations and determinations begin with estimation of the percentage of the five fundamental features of interest in C³I—IPG [6]. By granting analyzed percentages (in this case the Responsiveness and Opportunity of the system are of 40 % each), one can opt for which of the C³I—IPG system features are more relevant in the activities of obtaining and keeping for use of IFD.

In turn, the 5 fundamental features are split by the percentages the associated services hold within each. Hence the percentages of the services are estimated so that the sum of all to be exactly 100 % (for example, within Dispersion, "decentralization" can be rated at 40 %, the same as the "independence" of the leader system also at 40 %, leaving 20 % for "flexibility").

At the end of estimates must be granted percentages appropriate to Human and/or Technique influence within each service, so that the sum of these two influences to be 100 % within each service.

In order to establish a clear priority for the order in which starts and will run the entire operation/activity of monitoring the C³I—IPG integrated system, measurements determined by laborious calculations are used.

Because of their discrete probability distribution, reengineering processes can be considered Poisson distribution processes whose intensity can be determined [7].

Therefore success rate, due to high adaptability is known from the beginning, so that the system is flexible enough to allow access to data collected by advanced sensors. Also it should be designed to maintain safe levels of functionality. In addition, decision-makers, employers and organizers focus on the opportunity to develop new technologies to strengthen the C^3I—IPG as well as security (encryption) of Internet protocols, advanced firewalls, intrusion detection systems, even improved biometric systems.

Given the fact that connections will be determined by the actual circumstances or due to the need to use shared information, not only by a decision-maker, employer, organizer, but also by individuals, perhaps even by local government, it can appreciate that the project exceeds the IPG field as interest and importance.

By interpreting the results of these measurements, as can be seen from Fig. 3 *Associated Services, Connectivity*, is the most predominant associated service, followed closely by Adaptability and Early warning, continuing with Direction/monitoring, Execution time, Data Transformation, Relevancy, etc.

Based on these results, the characteristics Responsiveness and Opportunity become defining for the system and Dispersion and Mobility are less preponderant even to Invulnerability.

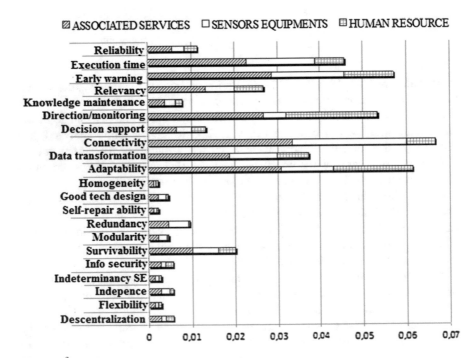

Fig. 3 C^3I—IPG priorities [6]

The diagrams from Figs. 4 and 5 show the same classification for establishing priorities for the order in which the entire renewal will start and run—actually the integrated system reengineering—in that defining are those characteristics that comprise over 80 % of the system prevalence: Responsiveness, Opportunity and Invulnerability.

As shown in Figs. 4, 5 and 6, Poisson distribution can be applied to electrical networks with a large number of possible events, each of which is rare. How many such events will occur during a fixed time interval, remain a problem to be solved after extensive monitoring activities of such networks. However, under the correct conditions, this is a random number with a Poisson distribution. Taking account of all presented before, a classification that reveal monitoring process is proposed with simultaneously two components of system in Table 2.

The next step is to analyze each type of equipment, judging by its usefulness in the complexity of C^{3}I-IPG system. Functional technical resources are useful in predicting situations, in conceptualizing methods to perform activities and include control centers (usually automated), communications (usually electronic), staff prepared and organized to operate technical resources (usually dependent on technology) as well as a planning specific and unique only to C^{3}I-IPG.

For example, it is considered that a communication terminal with non-terrestrial, satellite interface is a very special solution to meet the connectivity criterion. To ensure the early warning, are also necessary intelligent terminals with low

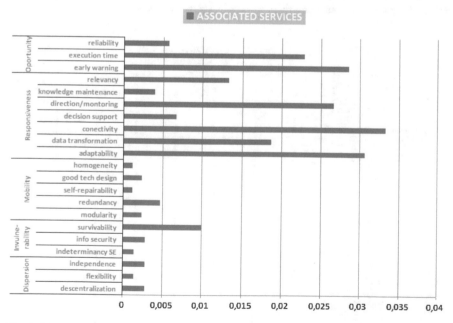

Fig. 4 Diagram representing ponderency associated services

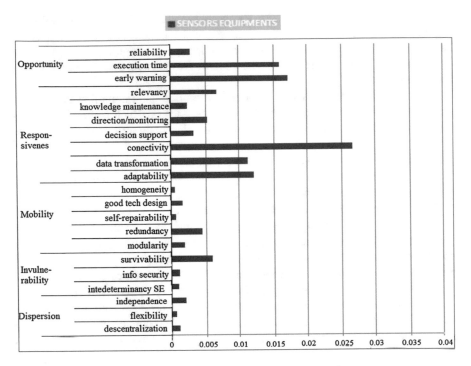

Fig. 5 Diagram representing priority of sensors or equipment renewal

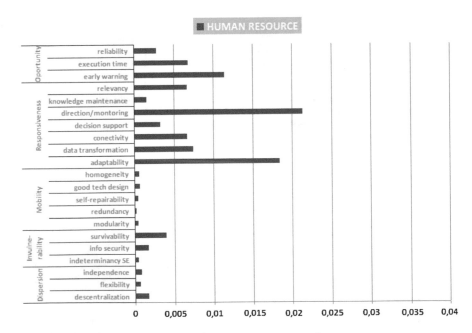

Fig. 6 Diagram representing priority of human resource renewal

Table 2 C^{3}I—IPG priorities

Priorities of monitoring the logistics of C^{3}I—IPG		
	Monitoring equipment sensors	Monitoring human resource
1	Connectivity	Direction/monitoring
2	Early warning	Adaptability
3	Execution time	Early warning
4	Adaptability	Data transformation
5	Data transformation	Execution time
6	Relevance	Relevance
7	Direction/monitoring	Connectivity
8	Decision assistance	FID viability
9	Confidence degree	Decision assistance
10	Database storage	Confidence degree
11	FID viability	Information security

transmission times and extremely reliable and Execution time must be supplied with advanced computers.

The most difficult question occurs about how much budget is needed. Some calculations show that a radical improvement of the system monitoring can be achieved, even with small budgets. But by using effective equipment or structural elements, which in addition to the fact that they respond to multiple criteria elements (reliability, universality, friendly interfaces, continuity of the standards and so on), have very low productive, and replacement costs.

This is the case of electro-optical sensors of measuring, controlling and capturing various physical quantities (temperature, pressure, humidity, electric field, and so on, conferring the possibility that the information to be forwarded using a determined period, which can oscillate between "real time" and "on demand", adapt to desired communications versions, electrical signals, IT network, intra/inter/ether net, are fast, without delays giving sufficient warning time, ensure a small execution time by synchronization or simultaneity of submissions [8–10].

4 Risk Analisys

There are several categories of risks envisaged, which could occur in the implementation and development of such re-designed networks and the most important ones are:

1. Market risk—MR—refers to how the "product" offered by the re-engineering process is received by the market. The manifestation of this risk can be seen in the variability of volume and/or price of the product. The new product does not

meet expectations in terms of marketing. This risk can be addressed by conducting a market survey aimed at time-dependent features of the products offered along the implementation and a market survey to know the specific features of competition, regarding parameters of other intelligent networks.

2. Technological risk—TR—may imply, in some cases limit that the technological "idea" is not correct and produces unwanted effects. The manufacturing system does not ensure sufficient productivity in order to be able to re-engineering large networks. This risk can be addressed by re-designing some technological operations and ensuring elasticity of the technological process to allow adjustments and changes.

3. Production risk—PR—it is generated by technological and organizational dysfunctions within the manufacturing activities. This category includes scrap, excess of the planned consumption. Operating costs are higher than those budgeted. This risk can be addressed by increasing the cost control and reviewing some technical and technological specifications.

4. Economic risk—ER—The risk is caused by cyclical developments in the economic environment that may affect the entire process of re-engineering networks in a certain area or at a global level.

The most common risks in this category are:

- Foreign exchange risk;
- The risk of increased cost of production;
- Investment risk.

The purchase prices of the equipment may increase because of the exchange rate variation. Any potential purchase agreements must provide for the trend of the exchange rate, in national currency, according to economic forecasts. Production costs may increase due to inflation. Measures to be taken refer to the introduction of a covering margin to the negotiated price, introduction in the agreement of a review or price adjustment clause.

5. Social risk—SR—it may occur due to poor relationships between the design team, or a failure to motivate the team members or even lack of skills in the field for the team members. It is manifested by poor communication and lack of an action plan in terms of human resources involved in the project. It can be mitigated through regular briefings and assessments of the project team members on the project status. In addition, it can be reduced by making the connection between results and wages, by establishing in an early planning stage of the project the level of involvement of each member, responsibilities, working time and the results expected from each one and the overall results.

6. Legal risk—LR—it arises from effect of domestic laws on the project activities. Thus, we can talk about risk of loss or destruction of goods purchased, risk of payment of additional taxes, risk of economic penalties, risk of misreporting.

5 Risks Management and Mitigation

Reengineering electrical networks and transforming them into future intelligent networks can be a risky enterprise. So, it important to expand the risk management process in any investment project of this kind, dealing both with monitoring and identification of new risks of the project.

Management practice demonstrates that risk monitoring and control are carried out while monitoring the project activities. Either is about new generations of sensors that need to be acquired, or data loggers that save and transfer monitoring parameters, they all imply a risk of a certain type. Risk monitoring and control strategies include identifying the extent of materialization of risk based on probability calculations, their registration and regular inspection of their development based on routine checks.

Construction of such a monitoring system must take into account the following aspects:

- The project manager must ensure that there is a person responsible for each category of risks;
- Composition of the risk recording system must consider risk classification based on the severity and probability of occurrence. We also have to highlight the importance of achieving a risk classification to identify response strategies in due time;
- Development of a system of continuous updating of the data included in the structure of the management system, based on routine checks, in order to observe the trend of risk factors in different phases of progress of the project.

Achievement of such a monitoring system, leads in a re-engineering project to avoid delays in the work projected, not exceeding the allocated budget and obtaining high quality results. Another component of the managerial work is the development of a strategy to identify continuously new categories of risk that may occur within the project, in order to prevent or mitigate them.

Monitoring for achieving objectives set for the work batches and of each individual activity conducted within these batches, are reference points for the identification of emerging risks that may arise along the way.

6 Conclusion

Analyses over the efficiency of reengineering processes in electrical independent networks are difficult to be performed, because of multitude of terms involved. Many of those terms have difficult definition and mathematical representations. Also, reengineering process becomes a priority, if alternative energies are to be delivered into the network, so efficiency is often placed in secondary positions.

Cost estimation remain a problem and covering these costs is not an easy option. But without a mathematical model that can estimate at least the level of costs, reengineering remain just a challenge, not an implementation.

The C^3I—IPG system monitored, can give an idea about the needed resources, using control, planning, directing, coordinating technical and human resources (equipment, communications, facilities and procedures) in achieving one, two or simultaneously three of the following goals:

- Optimization of energy consumption;
- Response capacity in different critical events over the network;
- Better monitoring the quality of energy distribution;
- Readiness of networks to absorb future energy sources.

Also, re-engineering process of a network imply the apparition of many risks, both technological and economical, so it is required to set up a system that is able to survey the whole process in order to find better methods to counter the influence of these risks.

References

1. Vintea, A., Schiopu, P., Ghita, O.: Optimal model for efficiency of reengineering the independent power grids. In: Proceedings of the 2014 6th International Conference on Electronics, Computers and Artificial Intelligence (ECAI), pp. 51–54. Bucharest, (2014). doi:10.1109/ECAI.2014.7090223
2. Boisrobert, J., Coutaz, L.: Materiaux Optiques, Photonique et Systemes. Universite Paul Verlaine—Metz, France (2004)
3. Hammer, M., Champz, J.: Reengineering the Corporation. Harper Collins Books, New York, NY (1993)
4. Carabulea, A.: "Industrial engineering course" (RO: "Tratat de inginerie industrială"). University Politehnica of Bucharest, Romania (2006)
5. Vintea, D.: The projection of communications system targeting the preservation of the statistic and informational flux of decision. In: International and National Conference on Sustainable Industrial Systems, pp. 52–60. University Politehnica of Bucharest, 2005
6. Bjorklund, R.C.: The Dollars and Sense of Command and Control, pp. .56–66, 201–218. Institute for National Strategic Studies Press, Washington, DC (2005)
7. Haight, F.A.: Handbook of the Poisson Distribution. John Wiley & Sons, New York (1967)
8. Grigorescu, S.D., Ghita, O.M., Neacsu, P.: "Virtual and distributed instrumentation" (RO: "Instrumentaţie virtuală şi distribuită"). Electra Publisher, Bucharest (2006)
9. IEEE Std 1100-1999: IEEE Recommended Practice for power and Grounding Electronic Equipment. IEEE Press (1999)
10. Agustoni, A., Borioli, E., Ferrari, P., Mariscotti, A., Picco, E., Pinceti, P., Simioli, G.: LV DC networks for distributed energy resources, In: Proceedings of CIGRE. Athens, 2005

Innovative Fuzzy Approach on Analyzing Industrial Control Systems Security

Emil Pricop, Sanda Florentina Mihalache and Jaouhar Fattahi

Abstract Industrial control systems are now very important components of the critical infrastructures. The security threats upon industrial control systems endangers the proper functioning of energy production facilities, the power grid, water production and distribution, chemical and petrochemical plants, food production facilities. This chapter presents an overview of potential and registered threats of industrial control systems. A risk assessment on this matter it is of high importance and the chapter has as focus an innovative fuzzy based approach on industrial control systems security. Two fuzzy based models are introduced, one for the attacker profile and one for attack success rate estimation. Finally, an industrial case study is presented with conclusions to proposed models limitations and challenges.

1 Introduction

Industrial control systems are technical systems that are used for monitoring and regulate a process in order to achieve an objective without any direct human intervention. In the last decade industrial control systems (ICS) become very important components of critical infrastructures.

E. Pricop (✉) · S.F. Mihalache
Department of Automatic Control, Computers and Electronics,
Petroleum-Gas University of Ploieşti, 39 Bucuresti Blvd., Ploieşti,
Prahova 100680, Romania
e-mail: emil.pricop@upg-ploiesti.ro

S.F. Mihalache
e-mail: sfrancu@upg-ploiesti.ro

J. Fattahi
Département d'informatique et de génie logiciel, Pavillon Adrien-Pouliot,
Université Laval, 1065, Av. de la Médecine, Québec, QC G1V 0A6, Canada
e-mail: jaouhar.fattahi.1@ulaval.ca

© Springer International Publishing Switzerland 2016
E. Pricop and G. Stamatescu (eds.), *Recent Advances in Systems Safety and Security*,
Studies in Systems, Decision and Control 62, DOI 10.1007/978-3-319-32525-5_13

Critical infrastructures are entities with a vital role in assuring the correct functioning of economic, social and informational processes [1]. According to US Department of Homeland Security definition critical infrastructures are all the systems whose deactivation endangers national security, economy, safety and public health [2]. It is obvious that energy production facilities, the power grid, water production and distribution, chemical and petrochemical plants, food production facilities, military and civilian communication systems are all national critical infrastructures.

The functioning of the energy production and distribution facilities, refineries, petrochemical plants and general factories is based on various interconnected automated control systems. The simplest control system has mainly three components: a controller, a transducer and an actuator, as shown in Fig. 1.

The controller has two inputs—the set point, the desired process value, and the feedback, the measured process value. The command signal is computed by the controller using a predefined algorithm and taking into account the two inputs specified above. The command signal is transmitted to actuator that directly action on the controlled process. It is obvious that informational signals are exchanged between control system components. The information can be manipulated in a simple cyber attack, making the system unable to reach its goal or, worse scenario, leading to accidents with severe impact on humans and environment [3].

The example presented above shows that even very basic control systems can be affected by security issues. The critical infrastructures are not operated by such simple systems, but by very complex ones often connected by specific industrial network protocols. The control technologies migrated toward communication and flexible technologies using commercial solutions and open protocols. This approach permitted the implementation of highly connected systems with remote control and supervision. In this way there were obtained an increase in efficiency and cost reduction but the systems became vulnerable to various threats and attacks [4].

ICS-CERT states in [5] that in the fiscal year 2014 a total number of 245 security incidents on industrial control systems were reported. The report [6] shows the evolution of the security incidents numbers from 140 in 2011 to 257 in 2013.

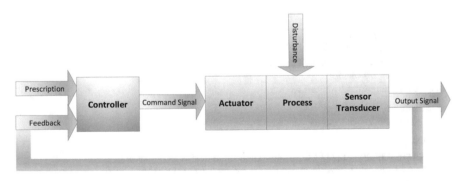

Fig. 1 Control system block diagram [3]

There should be mentioned also an increase in the severity and frequency of attacks targeting critical infrastructures. Only two examples are represented by Stuxnet Worm in 2010 and Flame in 2012. The result should be an increased interest of the academic community, governmental institutions and industrial world in correctly assessing the security risks associated with ICS and developing new methods and techniques for protecting the industrial infrastructures.

In this chapter the authors will present some methods for assessing the security risks associated with an industrial control system using fuzzy methods. This work represents an extension of two papers presented at the 2nd and the 3rd International Workshop on Systems Safety and Security—IWSSS in 2014 and 2015: *"Assessing the security risks of a wireless sensor network from a gas compressor station*[1]*"* and *"Fuzzy approach on modelling cyber attacks patterns on data transfer in industrial control systems*[2]*"* [7, 8].

This chapter is structured as follows. After a short introduction in the usage of fuzzy modelling for control systems security we will define the attacker profile. Each attacker profile has an associated score obtained by using a FIS based on attacker attributes—experience, knowledge, resources and motivation. A method to estimate the attack scenario success rate is presented in Sect. 4. The last section of this chapter consists in an industrial case-study, analyzing the security issues associated with a node located in the wireless sensors network from a gas compressor station.

2 Fuzzy Modelling Applied in Control Systems Security

The fuzzy inference systems (FIS) are an appropriate solution to describe into a linguistic manner the behavior of a process [9–14]. When there is vagueness in describing a process behavior, the FIS solution provides good results. The FIS approach is more suitable than conventional modelling methods in the case of nonlinear processes, if information about process nonlinearities is available [15, 16]. In order to model in FIS approach, the process must allow its behavior to be described by fuzzy IF THEN rules.

The architecture of a general fuzzy inference system is presented in the Fig. 2. A fuzzy inference system is formed by 5 functional blocks [17, 18]:

- the fuzzification pre-processing block: at this phase (the premise part) the crisp value of the input is transformed into a set of matching degrees to membership functions describing the fuzzified input;
- the defuzzification post-processing block: at this phase (the consequence part) the fuzzy set resulted from rule inference mechanism is transformed into a crisp value;

[1]DOI: 10.1109/ECAI.2014.7090209.
[2]DOI: 10.1109/ECAI.2015.7301200.

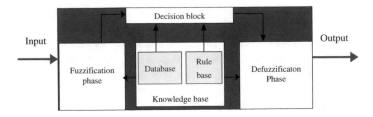

Fig. 2 Fuzzy inference system architecture [8]

- the database block: the storage unit for membership functions parameters for both input and output fuzzy variables (model's tuning parameters);
- the rule base block: the storage unit for the IF-THEN rules that describe system behavior (Mamdani-M or Takagi-Sugeno TS type);
- the decision block: the inference mechanism is applied to the existing rules.

The mathematical modelling of the process behavior usually begins with finding the proper membership functions for input and output variables (first set of tuning parameters of the model—the proper shape for membership functions: triangular, bell shaped, Gaussian, trapezoidal et al.). After choosing the proper shape, the tuning parameters for chosen membership functions are found using trial and error methods (second tuning parameters of the model). This phase is followed by the construction of the fuzzy rules (M or TS type, [19]). The IF-THEN rules are the result of observations and experience of an expert in process field. The last phase is applying the operations of inference to the fuzzy rules and then establishing the crisp output value and the modelling result.

In the following sections is proposed an innovative fuzzy approach on modelling the cyber attacks patterns on industrial control systems. In authors opinion the cyber attacks can be described with attributes such as attacker profile, protection degree of attacked unit, importance of the attacked unit (critical factor) and restore costs (impact factor).

3   Attacker Profile

In order to perform a correct security assessment on a given ICS, it is necessary to understand that an attack is characterized by three factors: the attack driver, the attack methodology and the system vulnerabilities/protection means. Each of these factors will be described in the following sections of this chapter.

The attack driver represents the entity that conducts a cyber attack targeting an industrial control system in order to reach an objective. The attacker could be an individual or a group such as a terrorist organization or a cyber-army. The objective of the attack can range from simply proud or recognition as a hacker to financial winnings or to accomplishing cyberwar missions.

Each attacker entity is characterized by a specific profile that can be quantified using an attacker profile score, as shown in [7, 8]. The attacker profile can be constructed as a semantic description of attacker abilities and behaviors, but this approach would not be effective for rapid and efficient assessment scenarios. The authors experience allows proposing a method to quantify the attacker profile based on three parameters: skills, technical resources and motivation. The parameters are described in the following paragraphs.

Attacker skills represent all the knowledge the individual or group acquired and can use in order to initiate and deploy a given type of attack. Those skills vary from a very low level, for example usage of a predefined tool to perform an attack, to a high level of computer and industrial process control knowledge. This parameter includes not only hacking and computer knowledge but also the awareness on the target system architecture, components and the existing security measures. It is well established that the attacker skills relay on his education level in the field of protocol specifications, hacking, computer engineering, programming and it is also dependent on the attacker experience. In the authors perspective this is the most objective approach. We assume that attacker skills could be affected by emotions, but their impact the reduction will not be significant in the context of this research.

Technical resources level is a key parameter for characterizing the attacker profile. It shows the level of hardware and software resources an attacker can access and use in order to deploy a given type of attack. The attack complexity is directly proportional to the resources level available, for example deploying a DDoS[3] attack requires a large network of compromised hosts (botnet), while a simple discovery attack might need only a computer and an IP scanning/sniffing software.

Motivation, the third parameter, is the most complicated to quantify. It shows how determined is an entity to deploy an attack. This parameter is very dynamic; it can change in various contexts. Motivation can be given by financial winnings, reputation, fame and even by religious or cultural beliefs, in case of terrorist attacks. The higher the motivation is, so the attack has a bigger success rate.

These three parameters that characterize the attacker profiles are used as the inputs for an FIS—fuzzy inference system. The output of this FIS is the attacker profile score, representing the attacker capabilities taking into consideration all the parameters presented in the previous paragraphs.

The construction and parametrization of the FIS was presented first in [8]. Three membership functions are used as granularity of model description for each input, resulting 27 fuzzy rules. We consider that this number of rules is adequate to fully represent the attacker profile. Increasing the number of membership functions (MF) for each input generates a significant growth of fuzzy rules number (MF number of inputs), with little consequence on the quality of the model versus computational effort.

---

[3]DDoS—Distributed Denial of Service.

Table 1 Inputs description [8]

Input	Membership functions		
	Small	Medium	Big
Resources	trapmf	trimf	trapmf
	[−0.225 −0.025 0.1 0.5]	[0.3 0.6 0.9]	[0.7 0.9 1.06 1.26]
Knowledge	trimf	trimf	trimf
	[−0.4 0 0.5]	[0 0.5 1]	[0.5 1 1.4]
Motivation	trapmf	trimf	trapmf
	[−0.45 −0.05 0.1 0.4]	[0.2 0.5 0.8]	[0.6 0.95 1.05 1.45]

Table 1 shows the most important features of proposed membership functions (trimf—triangular, trapmf—trapezoidal) for proposed inputs. These inputs were established based on authors' experience.

The output is described by 5 triangular membership functions as depicted in Fig. 3.

The attacker profiles can be obtained by various combinations of the parameters presented above. Not all the combinations interest the security analysts due to the very little probability to appear in day to day situations. Based on the authors experience we have selected some likely to appear profiles shown in Table 2. In the following section we will provide a description of each well-established profile along with its profile score.

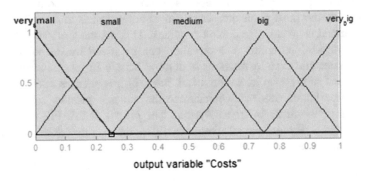

Fig. 3 Output of 5 triangular membership functions [8]

Table 2 Well-established attacker profiles

Attacker type	Main active rule				
	No	Resources	Knowledge	Motivation	Profile score
Script kiddie	1	Small	Small	Small	Very small
Disgruntled employee	8	Small	Big	Medium	Medium
Cyberterrorist	14	Medium	Medium	Medium	Medium
Cyber army	27	Big	Big	Big	Very big

The "script kiddie" attacker type corresponds to an unexperienced and unskilled individual. Often this type of attacker founds on the Internet various tools and exploits and try to use them in order to penetrate various protected systems. The motivation is gaining self-esteem and reputation. Their main targets are Web servers of well-known sites, e-mail services and rarely critical infrastructure services. This type of attacker is not very conscious on what is he really doing and the impact of their operation is not very significant. Unfortunately, the majority of cyber attacks are initiated by script kiddies.

In the case of script kiddie, the motivation level is small compared to the one of a well-paid attacker. When FIS is applied the resulted surface depends on resources and knowledge and it is presented in Fig. 4.

The script kiddie has not only little real knowledge on security issues but also small resources available in order to deploy an attack. His main resource is a software tool often found on the Internet without documentation. The hardware is frequently limited to the attacker own computer. In the particular case of small level of resources, the profile score depends mostly linearly on knowledge as depicted in Fig. 5.

Fig. 4 Attacker profile score in case of small motivation [8]

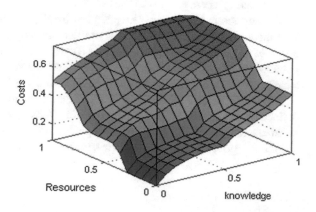

Fig. 5 Script kiddie—attacker profile score [8]

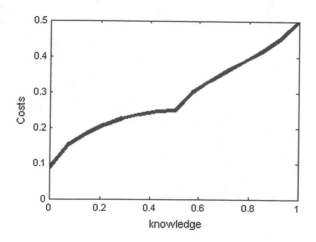

Disgruntled employees represent a special category of attacker posing a significant risk to any organization security. This attacker type is also included in the category of insider threats. An employee becomes disgruntled due to an unmet expectation or an unfortunate event, such as dismissal or promotion pass-over.

CERT insider threat database contains over 1000 incidents generated by insiders: sabotage, fraud or information theft. 33 of them were officially associated to disgruntled employees with an accent on industrial sabotage and data theft [20].

A good practice in industrial infrastructures is to monitor the control system activities, but it is very difficult to differentiate in real-time between a legitimate action and the one of a disgruntled employee. This drawback can be passed by using biometric user authentication, operation logging and multi-level access schemes.

It is obvious that disgruntled employee has a very good knowledge about the target system. He might have full access to documentation. The attack is often initiated from the interior of the industrial network. The main objective of the attacker is to get revenge by interrupting the functioning of the system, production sabotage and provoking economic losses. The attacker motivation is very dynamic and frequently decreases over time.

The surface in Fig. 6 was generated by the FIS when the motivation level is set to a medium level. In case of a disgruntled employee the resources available in order to deploy the desired attack is often lower, but it is compensated by knowledge. The profile score for this type of attacker is presented in Fig. 7. The attacker score is dependent on knowledge with a nonlinear saturation shape [8].

Another special attacker type is represented by terrorist entities. Terrorism is defined by the United States Department of Defense (DOD) as "the unlawful use of, or threatened use of, force or violence against individuals or property, to coerce and intimidate governments or societies, often to achieve political, religious or

Fig. 6 Attacker score in case of medium motivation [8]

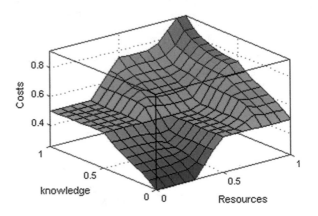

Fig. 7 Disgruntled
employee—attacker profile
score [8]

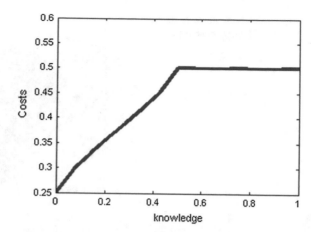

ideological objectives." [21]. A special case is represented by cyberterrorism, which can be referred as terrorist activities carried out in the virtual world. Computers and software components are valuable weapons used by such kind of attackers.

From our perspective a cyberterrorist is an individual or an organization with a decent level of resources available and with a strong motivation to interrupt the functioning of critical infrastructures. The attack, if successful, can have a big impact on economy and can spread terror on civilian population.

The cyberterrorist motivation is of medium level, but it is stronger than the one of a disgruntled employee. Their motivation consists in financial gain or satisfying an ideological duty. The resources and knowledge are often of medium level.

Cyber army or cyber warriors are two special attacker profiles characterized by a maximum of resources, knowledge and motivation. The cyberterrorist can evolve in a cyber army if enough funding and professional knowledge are obtained.

The cyber army is implicated in operations specific to a non-conventional electronic war. The objective is to break the functioning or even to destroy national critical infrastructures such as petrochemical plants, power grids or water distribution systems, in order to gain advantage from an enemy. The attacker has access to a high level of resources consisting in state of the art hardware and software. Often there are developers hired to write powerful software tools to conduct attacks. The motivation for the attack is estimated to be high, sometimes even very high. It is well sustained ideologically and financially. In this situation the surface resulted from FIS is presented in Fig. 8.

The cyber army attacker profile score is presented in Fig. 9. The resources level is set to big. The graph shows that the dependence from knowledge is not linear.

The above simulation results made in MATLAB shows the importance of knowledge factor in establishing the scores associated with attacker profile.

Fig. 8 Attacker profile score
in care of big motivation [8]

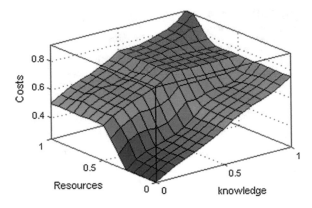

Fig. 9 The cyber warrior—
attacker profile score [8]

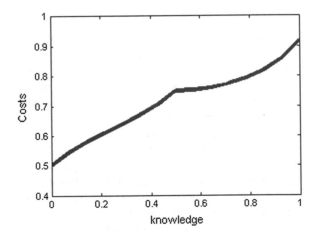

4 Attack Success Rate Estimation

The attacker profile score was defined in the previous section of this chapter. This
parameter is used now to introduce the attack success rate estimation. The success
degree of an attack depends on the type of attacker, existing system vulnerabilities
and the security measures taken to protect the infrastructure.

In this section we present the FIS used to estimate the attack success rate. The
proposed input-output structure of the model is presented in Fig. 10.

The vulnerabilities and the protection degree are described by three membership
functions (small, medium and big) as in Fig. 11.

The output of the model is described by 5 membership functions as in Fig. 12.

A totally successful attack can be defined as making the target system
unavailable for a given duration, commonly at least 5 min. Not all the attacks can

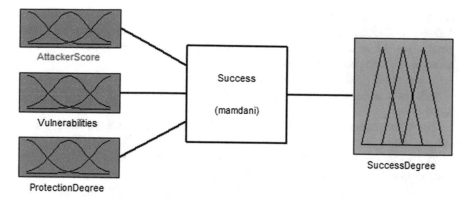

Fig. 10 Proposed input-output representation of the fuzzy model

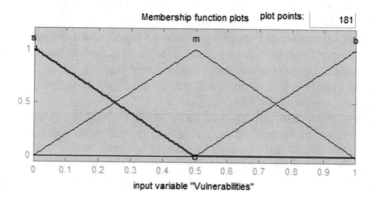

Fig. 11 The fuzzy representation of input parameter—vulnerabilities level

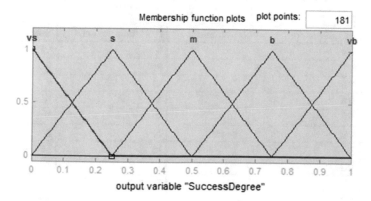

Fig. 12 Output of the FIS model

reach this goal so there is the need to define a measure of the attack effect. We define it as attack success ratio and it depends, as presented in the previous paragraphs, on the attacker profile score, vulnerabilities level and security measures taken in order to protect the system.

The attacker profile score was defined in the previous section of this chapter. It reveals the attacker abilities, so the complexity of the attack is proportional to this score.

The vulnerabilities level shows how critical are the known security breaches and vulnerabilities that are not mitigated by countermeasures.

The system protection level is an intrinsic system characteristic related to the number and the quality of countermeasures taken in order to secure the system.

Two inputs (vulnerabilities and protection level) are described by three membership functions; one input (attacker profile score) has five membership functions, in both cases with triangular shape. This yields to 45 fuzzy rules Mamdani type.

The fuzzy rules corresponding to the three inputs can be used to describe cyber attack pattern in industrial control systems. In Table 3 is presented a selection of the 45 resulted rules (VS-very small, S-small, M-medium, B-big, VB-very big).

The rules are constructed according to authors experience. For example, rule 15 represents a very possible to happen attack, upon an unprotected target deployed by a skilled attacker so the success rate would be very big.

Rule 35 specify a successful attack run by an expert attacker upon a highly protected target with small vulnerabilities.

Rule 6 describe an attack deployed by an attacker with a low profile score on a system with medium vulnerabilities and medium protection. This kind of attack will rather be unsuccessful and its success ratio will be small.

The success rate is directly proportional with attacker's profile score and the vulnerabilities and decreases with the protection degree, as presented in Fig. 13.

Table 3 Selection of proposed fuzzy rules for attack success rate estimation [7, 8]

No.	Main active rule			
	Attacker score	Vulnerabilities	Protection degree	Success degree
1	VS	S	S	VS
2	S	S	S	S
6	VS	M	M	S
10	VB	M	S	B
15	VB	B	S	VB
20	VB	S	M	B
27	S	B	M	M
35	VB	S	B	B
40	VB	M	B	B
45	VB	B	B	VB

Fig. 13 Attack success ratio in case of medium protection level [8]

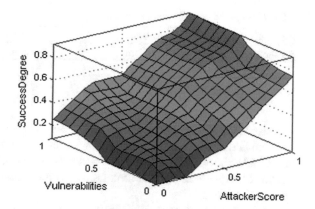

5 Industrial Case Study—Attack Success Rate Estimation for a Gas Compressor Station Wireless Sensor Node

In this section we will show how the fuzzy models presented above can be applied for estimating the success rate of an attack deployed on a wireless sensor node in a gas compressor station.

The studied gas compressor station was presented in detail in [7]. It is a component of the national gas transmission network and its function is to compress the gas to be stored for winter higher demands. The main goal is the compressor output to match the system demand. The critical parameters for the station are: the pressures and temperature for inlet manifold and outlet manifold, the level in oil tanks, the pressures for technical air, flowrate and pressure of utility fuel. These parameters are measured and transmitted to a control room by wire or wireless.

In the particular scope of our research the wireless pressure sensor on the inlet manifold is of special interest. It is a critical node of the wireless sensor network—WSN, in the station, since the inlet manifold comprise all treatments and preparing for gas compressor. This particular pressure transducer is of high interest because any malfunctioning can lead to plant shutdown (a higher/lower indication than the real pressure can lead to improper conditions in suction collector, compressor protection system can be misled). This is what makes this node highly important to attacks [7].

As presented in [7] we assume that the attacker goal is compressor station failure, by any means. There were identified two types of attacks that can lead to station failure:

- attacks regarding sensor availability;
- attacks on sensor data.

Table 4 Proposed costs for the given attack types

Attacker type	Knowledge level	Resources level	Attack cost
Sensor jamming	Medium	Medium	Medium
Denial of service	Medium	High	High
Data replay/injection	High	High	High

Sensor availability can be affected by a jamming and DoS[4] attacks. Jamming is the equivalent of sensor blocking, in order to make the system think it is unavailable/disconnected. Denial of Service (DoS) attack consist of making a large number of requests in a given time frame, so the sensor is overloaded and it can't reply to the request.

The exploits on sensor data consists in replay attacks and data injection. The sensor sends correct data, but the attacker intercepts and replaces it with invalid values and sends them to the correct destination. In this case the system behavior can become unstable and can lead easily to industrial accidents [7].

All these types of attacks have a big impact on the specified wireless sensor node in the compressor station. In order to be deployed each type of attack has a specific cost, representing the corresponding necessary level of resources and knowledge. In order to be used for with our fuzzy inference system the attack cost is specified by three values—low, medium and high. This data is presented in Table 4.

Sensor jamming is an attack that can be deployed by using a radio jammer to make the communication between the sensor and the access point impossible. The knowledge and resources levels needed are medium. The resources needed are the radio jamming installation on the given sensor frequency and also physical access to the node location. The attack cost shown in Table 4 is medium.

DoS attack is much complicated. Since there are many tools (scripts) available on-line for deploying such attacks we consider that DoS attack require only a medium level of knowledge. The resources needed are much higher that in case of simply jamming the sensor. An attacker will need access to some computers in order to deploy the attack. The resulting attack cost is high.

Data replay/data injection attacks are state of the art attacks. In this way an attacker can manipulate sensor data without being detected by the system owner/operator. The knowledge level required is high. Also it might be needed to build a node that can replace the target wireless sensor so the resources level is also high. Data injection is a complex type of attack, so its cost is high.

Taking into consideration the fact that the sensor node is physically isolated and it has low compute power and no intrinsic security measures the vulnerabilities level can be considered medium. The node runs a closed operating system and has implemented flood protection mechanisms so we can conclude that the node has a medium protection level according to Sect. 4 of this chapter.

---

[4]DoS—Denial of Service.

Table 5 Proposed costs for the given attack types

Attack type	Attacker profile	Attacker profile score	Vulnerabilities level	System protection level	Attack success rate
Compressor station failure due to inlet manifold pressure sensor unavailability	Script kiddie	Very small	Medium	Medium	Very small
	Disgruntled employee	Medium	Medium	Medium	Medium
	Cyberterrorist	Medium	Medium	Medium	Medium
	Cyber army	Very big	Medium	Medium	Big

In Table 5 are presented the results based on the FIS described in previous sections. We taken into consideration the previously defined attacker types correlated with their attacker profile scores for. The attack goal was making Compressor station failure due to inlet manifold pressure sensor unavailability, obtained by any type of possible attack (jamming, DoS or data injection).

Analyzing the results shown in Table 5 we can conclude that a successful attack is more likely to be deployed by a specialized attacker such an electronic or cyber army. A disgruntled employee or a terrorist with a medium attacker score may disrupt the functioning of the gas compressor station. It is very unlikely that a script kiddie would be able to affect in any way the wireless sensor network and to compromise the gas compressor station.

6 Conclusion

Fuzzy-based models use human expert knowledge to describe a system behaviour, functionality and characteristics. Data availability structured as knowledge is a condition to develop such models. Although there were intensively used from 1970s the main industrial applications were in control systems design. The security of industrial control systems is renewing its research directions due to the change in control systems connectivity and integrating strategies. The industrial wireless protocols have vulnerabilities that might be exploited by attackers.

The chapter presents an innovative fuzzy based approach to assess the risks of such industrial control systems that uses wireless control devices. Two fuzzy models are proposed: one that describes the attacker profile based on resources, knowledge and motivation and the other that calculates the success degree of an attack based on attacker score, vulnerabilities and protection degree. In the last section of this chapter an industrial case-study show how the proposed fuzzy models can be used to evaluate the security of a individual node in a wireless sensor network from a gas compressor station.

E. Pricop et al.

The research on modelling the industrial control systems security using a fuzzy approach should be continued in order to develop a security evaluation framework that can estimate the apparition risk for a given attack scenario and even to specify the needed security measures.

References

1. Alexandrescu G., Văduva, G.: Infrastructuri critice. Pericole, amenințări la adresa acestora. Sisteme de protecție. Editura Universității Naționale de Apărare "Carol I", București (2006)
2. Alexandrescu G., Văduva, G.: US Homeland Security Department Website. Critical Infrastructures. http://www.dhs.gov/what-critical-infrastructure. Accessed on 5 Nov 2015
3. Pricop E.: Security of industrial control system—an emerging issue in Romania national defense. "Mircea cel Bătrân" Naval Acad. Sci. Bull. **XVIII**(2), 142–147 (2015). ISSN: 1454-864X, Constanta
4. Kriaa, S., Pietre-Cambacedes, L., Bouissou, M., Halgand, Y.: A survey of approaches combining safety and security for industrial control systems. Reliab. Eng. Syst. Saf. **139**, 156–178 (July 2015). ISSN 0951-8320
5. Kriaa, S., Pietre-Cambacedes, L., Bouissou, M., Halgand, Y.: US Department of Homeland Security. ICS-CERT Monitor Sept 2014–Feb 2015, p. 1. https://ics-cert.us-cert.gov/sites/default/files/Monitors/ICS-CERT_Monitor_Sep2014-Feb2015.pdf
6. Kriaa, S., Pietre-Cambacedes, L., Bouissou, M., Halgand, Y.: US Department of Homeland Security. ICS-CERT Year in Review 2013, p. 16. https://ics-cert.us-cert.gov/sites/default/files/Annual_Reports/Year_In_Review_FY2013_Final.pdf
7. Pricop, E., Mihalache, S.F.: Assessing the security risks of a wireless sensor network from a gas compressor station. In: Proceedings of the 6th International Conference on Electronics, Computers and Artificial Intelligence (ECAI 2014), pp. 45–50, 23–25 Oct 2014. doi:10.1109/ECAI.2014.7090209 (IWSSS 2014 Volume)
8. Pricop, E., Mihalache, S.F.: Fuzzy approach on modelling cyber attacks patterns on data transfer in industrial control systems. In: Proceedings of the 7th International Conference on Electronics, Computers and Artificial Intelligence (ECAI 2015), pp. SSS-23–SSS-28, 25–27 June 2015. doi:10.1109/ECAI.2015.7301200 (IWSSS 2015 Volume)
9. Babuska, R.: Fuzzy Modelling for Control. Kluwer Academic Publishers (1998)
10. Driankov, D., Hellendoorn, H., Reinfrank, M.: An Introduction to Fuzzy Control, 2nd edn. Springer (1996). ISBN 978-3-662-03284-8
11. Holmblad, L.P., Østergaard, J.J.: The FLS application of fuzzy logic. Fuzzy Sets Syst. **70**, 135–146 (1995)
12. Jager, R.: Fuzzy logic in control. PhD Thesis, Technische Universiteit Delft (1995). ISBN 90-9008318-9
13. Jantzen, J.: Foundations of Fuzzy Control. Wiley (2007). ISBN 0-470-02963
14. Pedrycz, W.: Fuzzy Control and Fuzzy Systems, 2nd edn. Wiley (1993)
15. Russel S., Norvig P.: Artificial Intelligence—A Modern Approach. Prentice Hall (2010)
16. Takagi, T., Sugeno, M.: Fuzzy identification of systems and its applications to modelling and control. IEEE Trans. Syst. Man Cybern. **15**(1), 116–132 (1985)
17. Von Altrock, C.: Fuzzy Logic and Neuro-fuzzy Applications Explained. Prentice Hall (1995)
18. Wang, L.X.: A Course in Fuzzy Systems and Control, International edn. Prentice Hall (1997). ISBN 0-13-540882-2
19. Al-Jenani, N.A., Nounou, H., Nounou, M.: Fuzzy control of a CSTR process. 8th international symposium on mechatronics and its applications, ISMA (2012). ISBN-978-1-4673-0862-5

20. Al-Jenani, N.A., Nounou, H., Nounou, M.: Software Engineering Institute, Carnegie Mellon University, Insider Threat Blog. Handling Threats from Disgruntled Employees. https://insights.sei.cmu.edu/insider-threat/2015/07/handling-threats-from-disgruntled-employees.html. Access date: 20 Nov 2015
21. Al-Jenani, N.A., Nounou, H., Nounou, M.: US Department of Defense—DOD Dictionary of Military and Associated Terms http://www.dtic.mil/doctrine/dod_dictionary/. Access date: 23 Nov 2015

IPv6 Sensor Networks Modeling for Security and Communication Evaluation

Ionela Halcu, Grigore Stamatescu, Iulia Stamatescu
and Valentin Sgârciu

Abstract Ubiquitous monitoring and control in future internet-connected pervasive systems architectures rely increasingly on Wireless Sensor Networks (WSN) as enabling technology. This can be achieved through seamless integration along existing networking devices and systems, while taking account the data-driven nature and security constraints imposed upon such low power, efficient embedded networked platforms. We argue that only through a combined security and communication approach can the relevant goals be achieved, by considering security issues as an integral part of high level layers of the protocol stack. This chapter presents an extended analysis of security and communication constraints of IPv6-based wireless sensor networks at the network and transport layer. Leveraging the Contiki operating system for embedded sensing nodes along with the ContikiSec security layer and RPL, IPv6 Protocol for Low Power and Lossy Networks. The results of the analysis are discussed and functions of its modules are explained with focus on future reliable and large scale implementations.

Keywords Wireless sensor network · ContikiSec · Security · IPv6 · 6LoWPAN

I. Halcu (✉)
Intelligent Measurement Technologies and Transducers Laboratory, University
"Politehnica" of Bucharest, 313 Splaiul Independentei, 060042 Bucharest, Romania
e-mail: ionela@imtt.pub.ro

G. Stamatescu · I. Stamatescu · V. Sgârciu
Department of Automatic Control and Industrial Informatics, Faculty of Automatic
Control and Computers, University "Politehnica" of Bucharest,
313 Splaiul Independentei, 060042 Bucharest, Romania
e-mail: grigore.stamatescu@upb.ro

I. Stamatescu
e-mail: iulia.stamatescu@aii.pub.ro

V. Sgârciu
e-mail: vsgarciu@aii.pub.ro

© Springer International Publishing Switzerland 2016
E. Pricop and G. Stamatescu (eds.), *Recent Advances in Systems Safety and Security*,
Studies in Systems, Decision and Control 62, DOI 10.1007/978-3-319-32525-5_14

1 Introduction

Though far enough from the early visions of pervasive computing and "smart dust", wireless sensor networks have steadily evolved into a robust field of research and commercial applications as well. Sensor nodes have become embedded into the fabric of everyday life through deployments related to the future electric grid, smart cities and smart homes, mobile health, etc. Most current challenges are especially related to two major topics: security and communication. Questions directly associated to these topics are: "How can the integrity, confidentiality, and security related properties, of the collected and aggregated data be assured?" or "Which is the most efficient networking protocol under given specifications and performance bounds?" These are fundamentally linked to the constrained nature of the computing and communication resources available at the node level which limit the scale and complexity for new approaches. Thus, the need has arisen to port several security algorithms and mesh communication protocols to these low-cost platforms or to develop from the ground up specific solutions tailored to the desired application domain.

The usage of IPv6 communications in WSNs is becoming more popular since the development of 6LoWPAN by the IEEE 802.15.4 working group [1]. Although security represents an important issue in any IP-enabled communications, using IPv6 in WSNs on a large scale basis is currently an open issue.

Current tools and toolchains for prototyping and simulation range from low-level modeling of embedded microprocessors and radio transceivers to networking tools for generic simulation of network stacks with imposed constraints on available resources towards full system environments, usually associated to the two major platforms for WSN implementation, like TinyOS and Contiki OS [2]. Within this context, the task of the sensor network system designer is first to suitably choose among a range of available algorithms, protocols and platforms which build up a complete system under experimental analysis and validation using widely accepted hardware and software tools.

The main contribution of this work is to provide an experimental analysis for a combined security and communication approach in a 6LoWPAN WSN system design. We leverage the Contiki/COOJA simulation environment in order to gather and investigate data stemming from the TelosB/Tmote Sky mote platforms running the ContikiSec security layer on top of IPv6 wireless mesh communication. As far as we know this is the first implementation of ContikiSec on the TelosB/Tmote Sky mote platform. The results provided provide a solid background for further development of new algorithms and protocols for robust real world WSN deployments. Our main view is that through this type of approach the development time can be significantly reduced and bugs eliminated, preventing costly and sometimes inaccessible fixes for remote or critical applications, while at the same time building trust among future end-users of embedded networked technology.

The chapter is a revised and extended version of [3] and the rest of the chapter is structured as follows. In Sect. 2, we give an overview of the main benefits security has over wireless sensor networks, and we describe the security properties along with constraints and specification of IPv6 over sensor network technologies. Section 3 discusses related work in the field of wireless sensor network security with existing architectures and security technologies implemented in WSNs. Section 4 refers to the implementation of ContikiSec mechanism in a typical 6LoWPAN network. Simulation results using ubiquitous computing tools like the Contiki/COOJA for prototyping wireless sensor networks in a reference scenario are given in Sect. 5. Section 6 presents main conclusions and highlights future work directions.

2 Background

Wireless sensor networks (WSN) are becoming increasingly important in various application domains. Previous research [4] has shown the importance regarding real future systems aimed at high-resolution measurements of environmental parameters for indoor spaces. Interconnecting almost all heterogeneous devices into the Internet is the next step, to explore the full potential of these networks. Many devices are already connected to the Internet and transition to IPv6 becomes unavoidable as the number of connected devices is significantly expanding. Integrating WSNs into the Internet, however, requires certain features and specification to work, for example, adaptation of the respective link technology, specification of ad hoc networking, mobility management, handling the security issues and auto-configuration to support ad hoc deployment. Connecting WSNs to the Internet using IPv6 delivers further benefits such as scalability, providing support for network mobility and built-in auto-configuration features.

In this section, a brief description of WSN constraints is provided, along with the security properties that must be enabled in every WSN application. The features and specification of IPv6 communication over sensor network technologies are also provided.

2.1 Constraints in Wireless Sensor Networks

Wireless sensor networks are formed of a great number of resource-restricted sensor nodes. These nodes have different limitations such as processing, capacity storage, and communication bandwidth because of restricted power and size of the sensor nodes. Because of these restrictions, using the formal security techniques in WSNs is not always the best solution, it is essential to optimize the routine security algorithms.

Due to the wireless nature, a WSN is susceptible to threats and risks. For example, an attacker can compromise the integrity of a sensor network by several attacks such as: snooping, spamming, introducing wrong messages, destroy network resources, among others. WSN nodes broadcast their messages to the medium differently from wired networks.

For WSNs, the goal is to design, develop and implement IPv6-enabled sensor networks over the wireless environment, addressing requirements on the architecture and its functional blocks, along with the security problems that IPv6 networks implies.

2.2 Security Properties in Wireless Sensor Networks

The security properties that should be properly guaranteed by a WSN network are briefly described below.

- **Confidentiality** is the main property of every secure communication system. Confidentiality assures that information is kept secret from unauthorized third parties. The typical way to achieve confidentiality is by using symmetric cryptography for encryption with a secret shared key. The information that needs to be protected depends on the application.
- **Authenticity**. Data authenticity allows legitimate sensor nodes to be able to detect unauthorized messages and to reject them. An opponent is not limited to changes in data packets, he can also change the whole flow by introducing additional packages. The mean to achieve authenticity is to include a message authentication code (MAC) in every packet. A packet's MAC is calculate using a secret shared key, which can be the same one used for encryption. Moreover, MAC also offers integrity.
- **Integrity** is the guaranty that the package has not been modified during transmission. Data loss or errors can occur in a network not only in the presence of malicious nodes, but also due to difficult communication environment. Integrity can be achieved by including a message integrity code (MIC) or a cyclic redundancy checksum (CRC) in every packet. CRC or MIC is calculated by a cryptographic hash function that detects malicious attempts or accidental errors during packet transmissions.
- **Privacy** is a very important issue in terms of a network security. In most applications, the nodes transmit rather sensitive data, such as the distribution of keys, so it is very important to build a secure communication channel in a wireless sensor network. The node's public information, such as identities and public keys should also be encrypted as an extra security against traffic analysis attacks.
- **Availability** is a security requirement that must be ensured throughout the network, without affecting the operational structure of the network. By adjusting traditional encryption algorithms to fit in a wireless sensor network, additional costs are introduced. Several approaches have chosen to modify the code for its reuse as much as possible, or to simplify the algorithm, but all these approaches weaken the availability of a sensor and the entire network. Additional

calculations and communications consume additional energy. Moreover, as the communications grow, arise the risk of communication conflicts. All of these threatens the network availability.

- **Freshness.** Even if confidentiality and data integrity are assured, it is necessary to check data freshness of each transmitted message. Informal, data freshness suggests that the data are recent, and no old message was retransmitted. This requirement is particularly important when there are key exchange strategies imposed by network design. Typically, the key distribution must be done periodically. However, takes a long time for new keys to be propagated throughout the network, and in this case, it is easy for an attacker to inject reply messages. Also, the activity of a sensor can be interrupted if it is not aware of the key exchange. To solve these problems, a timer can be added to each data packet to ensure freshness.
- **Nonrepudiation** proves the source of a message. By authenticating the source, it demonstrates its identity. Nonrepudiation prevents the source from denying that it sent the message.

2.3 Specification of IPv6 over Sensor Network Technologies

The use of IPv6 for WSNs has been investigated for many years. One of the research main outcomes in this area is a compressed version of IPv6 for low-power devices, namely IPv6 over Low Power Wireless Personal Area Networks (6LoWPAN) [5]. It is a simple and efficient mechanism that allows to reduce the size of the IPv6 address for constrained devices, while enabling border routers to translate these compressed addresses into regular IPv6 addresses. The following features comes with integrating IPv6 into WSNs.

Ad Hoc Networking. In order to increase the reachable range within the sensor network, IPv6-enabled sensor networks are expected to form a hierarchical network in which IPv6 data packets are forwarded by the intermediate nodes towards the packet's destination. IPv6 provides end-to-end connectivity with a more distributed routing mechanism than IPv4. The IPv6 protocol makes routing more efficient and hierarchical by reducing the size and complexity of routing tables. Concerning the topology, WSNs could be dynamic, i.e. nodes may enter and leave the network or just move around. In various scenarios, the setup of WSNs is made in an ad hoc fashion. The initial routing configuration can be generated automatically. IPv6 defines several multicast addresses for the Internet auto-configuration procedures. IP multicast is particularly useful in the WSN environment. In large WSN deployments, it allows to distribute commands or set of data to all the sensor nodes on the Wireless Area Network (WAN).

Auto Configuration. WSN nodes need to be configured with several parameters (radio channel, channel check rate, RDC and MAC protocols, and IPv6 addresses) to make them ready for communication at the network layer. The configuration of the nodes can be made in various methods: manually, using human-machine interfaces (HMI), through the manufacturer provided default configuration or through the self-configuration mechanism (stateless mechanism). Sensor networks may consist of hundreds of nodes deployed in an ad hoc fashion, making manual configuration of nodes a difficult task. In contrast, auto configuration of IP addresses allows WSN nodes to become communication ready enabling drastic reduction of the configuration effort and cost for managing the system.

Mobility Management. In case of a mobile WSN, a dynamic point of attachment of the WSN to the Internet is required. For example, the sensor network is attached to a person (Body Area Network) or a vehicle that moves around. Moving networks between different geographic locations must support roaming, which is often handled by the subnet technology in use. For example, Wireless Local Area Networks (WLAN) support roaming between different access-points at the link layer and cellular networks allow a seamless roaming. However, when roaming takes place between different communications technologies it becomes difficult to resolve it at the link layer and has to be handled at the network layer. In order to keep running a session and maintain end-to-end connectivity in a mobile WSN, mobility management is required. Without the NAT functionality between the sensor network and the Internet, an IPv6 enabled sensor network provide strong features and solutions to support mobility of both end-nodes and routing nodes.

Security. Security is a major concern in every part of the Internet, covering areas like encryption, detection of intrusion, access control, authentication, authorization, integrity protection, prevention of Denial of Service (DoS) attacks etc. However, there are some limitations in merging security into WSNs because of the resource constraints concerning processing power, storage and network bandwidth. To overcome the security limitations, new features have been included in the design of IPv6. Among the features that support or improve security, we can mention: the introduction of IPsec, large addressing space and Neighbor Discovery.

3   Existing Security Architectures for WSNs

In recent years, many efforts have been focused on designing secure systems. New approaches must take into consideration two main problems related to security protocols: the overhead that protocol imposes to messages must be reduced and reasonable protection must be provided while limiting resource usage. In this section, we consider the security architectures currently implemented in the development of WSNs. We present the state-of-the-art of these architectures and evaluate them according to the requirements presented in Sect. 2.2.

3.1 *Authentication and Encryption Solutions*

TinySec [6] is link-layer security solution designed for the TinyOS [7] operating system. TinySec is designed to be as lightweight as possible and has minimal impact on WSNs performance, implementing security layer which adds very little overhead to original TinyOS packet exchange. TinySec addresses three main security requirements: confidentiality, integrity and authentication. TinySec support two modes of operation, TinySec-AE and TinySec-Auth. TinySec implements Skipjack as the default block-cipher, with an 80-bit key length and CBC-CS as the operation mode.

MiniSec [8] is another security layer for WSNs, inspired by TinySec. It offers confidentiality, authentication and reply protection. MiniSec supports two operating modes, depending on the type of communication being used, single-source or multi-source broadcast. MiniSec uses a special mode of the Skipjack block-cipher, Offset Codebook Mode (OCB), which provides authenticity and confidentiality.

FlexiSec [9] is an energy-efficient, configurable link layer security architecture for WSNs. It is intended to be responsive to the actual security demands of the applications. FlexiSec architecture is a configurable software-based security solution by adopting configurable block ciphers, configurable modes of operations, configurable MAC sizes and configurable replay protection mechanisms. The architecture is aimed to offer the optimal level of security at minimal overhead.

ContikiSec [10] is a link-layer security solution for wireless sensor networks introduced in 2009, and designed for the Contiki operating system. The design of ContikiSec is heavily influenced by TinySec, trying to be minimalistic and providing only basic security services. ContikiSec has been designed to balance low energy consumption and security with the need to comply with small memory footprint. ContikiSec provides confidentiality, integrity and authentication through its three security modes: *ContikiSec-Enc*, provides confidentiality only; *ContikiSec-Auth* provides integrity and authentication; *ContikiSec-AE* provides confidentiality, integrity and authentication. The source code of ContikiSec is based on security algorithms such as AES, CBC-CS, CMAC and OCB.

- *Offset Codebook Mode* (OCB) is a mode of operation that simultaneously provides confidentiality and authenticity, as it allows encryption and authentication in a single pass. Advantage of OCB is that instead of performing two operations (Encryption, MAC) for achieving confidentiality, integrity and authenticity, only one operation (authenticated encryption) is performed, which reduces time of computation and achieves the same result.
- *Cipher Block Chaining-Ciphertext Stealing* (CBC-CS) requires an initialization vector (IV), which is a random block of data, to achieve the confidentiality property.
- *Cipher-based Message Authentication Code* (CMAC) provides assurance of the authenticity and the integrity of binary data.

Similarly to TinySec, confidentiality is achieved using block-cipher in CBC-CS mode. Contiki uses AES instead of Skipjack, as it is proven to be more secure, although more computationally intensive. Integrity and authenticity in ContikiSec are very similar to TinySec and uses a 4 Byte CMAC (which is a variation of CBC-MAC). The frame format of each security mode operation is described in Sect. 4. Similarly to TinySec, ContikiSec does not define any key exchange mechanism. Instead a network-wide symmetric key is used, which makes ContikiSec vulnerable to attacks based on compromised sensor nodes.

3.2 Key Management Mechanisms

A special attention must be given to the key management field in a WSN. Key management and encryption are crucial in securing a wireless sensor network. Most traditional encryption techniques, however, are unsuitable for low-power devices such as wireless sensor networks. This is largely due to the fact that typical techniques are using asymmetric cryptography for key exchange (public-key cryptography—PKC). In this case, it is necessary to maintain two mathematically related keys, one of which shall be made public while the other is kept private. Such data is encrypted with the public key and decrypted with the private key. The problem of PKC in a wireless sensor network is that it is computationally intensive, speaking of individual nodes. However, there are also achievable solutions with a suitable selection of algorithms, related to key distribution [11–14].

In [7], a classification of key management schemes for WSNs has been made. The key establishment techniques applied in a certain WSN takes into consideration all the requirements, constraints and various characteristics of the employed network. The simplest key management technique is preloading the network with a *single wide-shared key*. A number of various key management schemes have been proposed, i.e. *pairwise key establishment, public key schemes* (PKC, ECC), *key predistribution schemes, hierarchical key management* etc.

Using a key management system requires extra processing in terms of computation and communication. It is important to analyze the performance of the sensor network as well as the key management scheme implemented, before discussing the various security techniques.

3.3 Contiki OS and μIPv6

Contiki [2] is an open source, highly portable, multi-tasking operating system for memory-efficient network-embedded systems and wireless sensor networks, led by the Swedish Institute for Computer Science (SICS) since 2004. Contiki targets low-power embedded microcontrollers which typically pose severe memory constraints. The software platform is been continually developed by both scientific community and commercial support, a significant advantage being that the platform has been ported to many WSN hardware platforms, such as TelosB/Sky, MicaZ, Raven, ESB, Sensinode, etc. The basic operating model of Contiki is of hybrid nature, relying on an event-driven kernel on top of which pre-emptive multi-threading operates in the form of an application library. Two significant advantages are offered by Contiki for the developer community. First, that C language is used for development, libraries and tool-chains, in comparison with TinyOS which makes use of the C dialect called nesC introducing an additional hurdle to adoption and development. Second, Contiki has already been blended in with proprietary technologies from many companies and is already found in products ranging from smart home appliances to intelligent electrical meters for future electrical networks [2]. The latter kind of applications are the ones driving the need to include security of processing and communication on constrained embedded platforms, providing relevance to this analysis. Contiki offers a standard to support IPv6 networking on low-power nodes: μIPv6 as a 6LoWPAN implementation. Figure 1 illustrates the high-level architecture of the Contiki system with μIPv6 stack included.

At the bottom layer, hardware support in the form of specific drivers is divided among CPU and platform, assuring flexibility. Medium access control (MAC) is carried out through Rime and at the upper-level μIPv6 implements IPv6 networking by a socket-like API for all UDP, TCP and ICMPv6 protocols. User applications make use of OS-provided low-level services for timing and pseudo-threading. This modular, simplified architecture enables efficient deployment of applications and adaptation across specific application domains. Through IPv6 it is possible to address every sensor node and access them from outside the network. In addition, μIPv6 has a small footprint and memory usage that allows it to run on most constrained platforms. A typical implementation has a code size approximately 11.5 kB and the RAM usage around 1.7 kB. Other key features of the IPv6 are the simplified header format, its improved support for extensions and options, and the

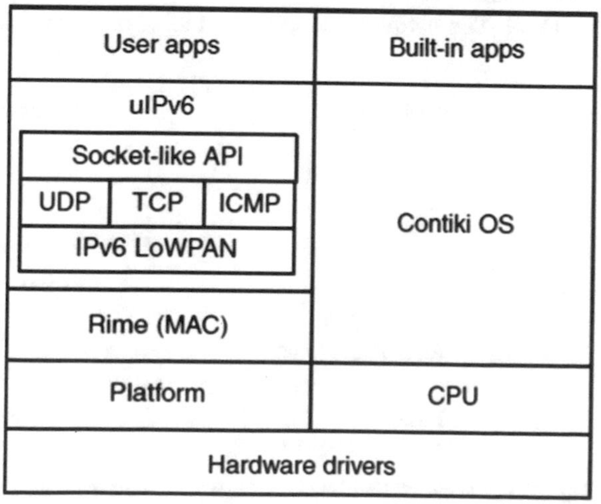

Fig. 1 The Contiki architecture with IPv6 enabled

QoS and security capabilities. In this paper, the μIPv6 stack was tested on a wireless sensor network simulation environment. Along with IPv6 communication, the issue of implementing security operations on constrained devices is also addressed.

4 Implementation of ContikiSec Mechanism

In order to evaluate the benefits of using Contiki's μIPv6 along with a security solution for a 6LoWPAN network, we implemented the ContikiSec mechanism [10] in a simulation environment. The network topology used for testing is presented in Fig. 2. Our selected platform is composed of Tmote Sky (TelosB) sensor nodes that run on top of the Contiki operating system. The main features of the hardware are described below. Features of the operating system were described in Sect. 3.3. As far as we know this is the first implementation of ContikiSec on the TelosB/Tmote Sky mote platform. The mechanism was initially designed and tested only for the MSB-430 platform [2]. Also, ContikiSec does not define any key establishment technique, as we intend to use.

The Tmote Sky is a hardware platform for wireless sensor networks. The Tmote Sky platform is a MSP430 F1611 based board with 802.15.4 compatible CC2420 2.4 GHz radio chip. The motes operates at 4.15 MHz, it has 10 kB of RAM and 48 kB of ROM. Moreover it is one of the first platforms to provide an open implementation of the 6LoWPAN adaptation layer and high-level protocols like RPL (IPv6 Routing Protocol for Low Power and Lossy Networks) [15], CoAP etc. Tmote Sky also supports energy optimization features such as auto-suspend, wake, and sleep mode.

The overall structure of Contiki's protocol stack with ContikiSec enabled is organized as in Fig. 2.

The CC2420 radio transceiver has security support based on IEEE 802.15.4 protocol. The hardware support enables a high level of security for a very low-cost

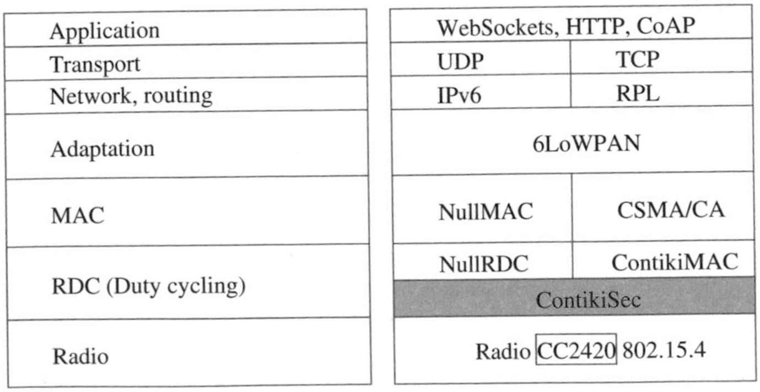

Application	WebSockets, HTTP, CoAP	
Transport	UDP	TCP
Network, routing	IPv6	RPL
Adaptation	6LoWPAN	
MAC	NullMAC	CSMA/CA
RDC (Duty cycling)	NullRDC	ContikiMAC
	ContikiSec	
Radio	Radio CC2420 802.15.4	

Fig. 2 The standard Contiki stack (*left*) and modified with *ContikiSec* security sublayer (*right*)

system. All security operations are performed within the transmit and receive of frames and are based on AES 128bit encryption. The latest specification version [16] describes three security modes of operations: counter mode (CTR) encryption/decryption, CBC-MAC authentication and CCM encryption plus authentication. The modality of key establishment used for encryption and authentication is decided at an upper level. By implementing ContikiSec as a new layer on top of the physical layer, the user has full flexibility in selecting the security options for each particular application.

Radio Duty Cycling (RDC) and Medium Access Control (MAC) protocols are an important part of the Contiki stack as they ultimately determine the power consumption of the nodes and their behavior when the network is congested. Contiki provides a set of RDC mechanisms, with various properties. The most commonly used are ContikiMAC, X-MAC, CX-MAC, LPP, and NullRDC. ContikiMAC is the default mechanism that provides a very good power efficiency, and is tailored for the 802.15.4 radio and the CC2420 radio transceiver. X-MAC is an older mechanism that does not provide the same power-efficiency as ContikiMAC but has less stringent timing requirements. CX-MAC (Compatibility X-MAC) is an implementation of X-MAC that has more relaxed timing than the default X-MAC and, therefore, works on a broader set of radios. LPP (Low-Power Probing) as a receiver-initiated RDC protocol. NullRDC is a "null" RDC layer that never switches the radio off and, therefore, can be used for testing or for comparison with the other RDC drivers. The MAC layer is responsible for avoiding collisions at the radio medium and retransmitting packets if there was a collision. Contiki provides two MAC mechanisms: CSMA (Carrier Sense Multiple Access) and NullMAC. CSMA is the default mechanism. The MAC layer receives incoming packets from the RDC layer and uses the RDC layer to transmit packets. If the RDC layer or the radio layer detects a radio collision, the MAC layer may retransmit the packet at a later point in time. The CSMA mechanism is currently the only MAC layer that retransmits packets if a collision is detected.

In Fig. 3 the default Contiki packet format for the CC2420 radio and the packet format for each security mode offered by ContikiSec, are presented. The ContikiSec packet formats are explained in more detail in [10].

The additional modifications of the Contiki packet format are highlighted as it can be seen in Fig. 3. Packet overhead occurs for both CBC-CS and OCB operation

PBL	S	H	PAYLOAD	C	T			Contiki packet
6	2	2	0..128	2	2			
PBL	S	H	IV	PAYLOAD	C	T		ContikiSec-Enc
6	2	2	2	0..128	2	2		
PBL	S	H	PAYLOAD	MAC	T			ContikiSec-Auth
6	2	2	0..128	4	2			
PBL	S	H	IV	PAYLOAD	MAC	T		ContikiSec-AE
6	2	2	2	0..128	4	2		

Fig. 3 ContikiSec packet formats compared to the Contiki packet format

Table 1 ContikiSec security modes

	ContikiSec-Enc	ContikiSec-Auth	ContikiSec-AE
Properties	Confidentiality	Integrity and authentication	Confidentiality, integrity and authentication
Operation mode	CBC-CS	CMAC	OCB
Time (μs)	99	99	122
RAM (kB)	0.21	0.21	0.24
Code size (kB)	39.74	41.54	45.5

*The cipher used is AES with 128-bit key size

modes. Differences in lengths can also be noticed. In addition, a 2-byte CRC byte is auto-appended to each outbound packet by the CC2420 radio hardware itself. However, the checksum is designed to detect only accidental modifications of the data, whereas a MAC also detects intentional unauthorized modifications of the data. ContikiSec-Auth removes the checksum field and provides both authentication and integrity by including a MAC field.

The three supported security modes are the main object of this analysis: *ContikiSec-Enc, ContikiSec-Auth* and *ContikiSec-AE*. These offer confidentiality-only, authentication-only and authentication with encryption security features. For a real deployment this allows flexibility in tailoring the security level and processing and communication requirements to the application constraints. In Table 1 are defined the three security modes of ContikiSec and different parameters associated to them.

The security operations (encryption, decryption and authentication) are done upon sending and receiving data packets. In Table 1 it is presented the timing and RAM usage for each operation during the transmission or receiving a packet by a node, and the code size in terms of memory footprint on a regular node. One of the characteristics of sensor network applications that must be taken into account when choosing a security solution is the energy consumption of sensor nodes. To analyze the energy overhead for each security mode of ContikiSec, we used a software-based solution to estimate the energy consumption of a sensor node [17].

As shown in Fig. 4, the energy consumption of Contiki in default mode ranges from 80 to 170 μJ for packets with payloads of 20–40 bytes. With security applied, ContikiSec-Enc and ContikiSec-Auth have the same energy consumption because both incur a 2-byte overhead. Finally, ContikiSec-AE has the highest energy requirement, consuming around 15 % more than Contiki in default mode for data messages with payloads of 40 bytes.

Fig. 4 Energy consumption of ContikiSec security modes

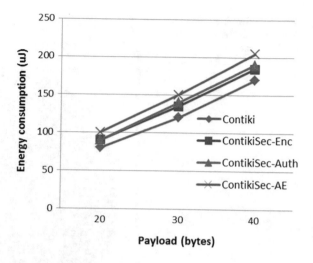

5 Evaluation and Simulation Results

In order to simulate and implement 6LoWPAN/IPv6 protocol, we used the Contiki dedicated simulation environment for wireless sensor networks called COOJA [18]. The network simulator is able to emulate the Tmote Sky sensor platform, meaning that the same code executed by the simulator is exactly the same firmware that runs on physical nodes.

Due to the resource limitations of these types of networks, it is essential to measure the network performances in terms of efficiency, memory, encryption speed, and energy consumption. In Fig. 5 it is presented the 6LoWPAN network used for testing.

Figure 5 illustrates our test network with 15 Sky Motes. In the testing scenario, we use 14 sensor nodes (senders) that send packets toward the sink node. In fact, all the application traffic is destined to the node 1 (sink), since this mote acts as a router to forward the data to Internet. The sink node is running the UDP protocol over RPL. The IPv6 packets are sent over 802.15.4 network with 6LoWPAN header compression. The border router has the role to setup the IPv6 prefix of the network and to generate the RPL routing tree. The general parameters for the simulations are presented in Table 2.

The wireless channel used for simulation is Unit Disk Graph Medium (UDGM)—Distance Loss, in which two nodes can communicate to each other only if they are within the same transmission range. The interference range represented by the gray circle (see Fig. 5), is the area where radio interference from current node exists. The network topology is selected as a mesh topology. The nodes have the configured route up to 7 hops, as we can see in Fig. 6. The links of the employed routes have a rather high success rate in case of other ongoing trans-missions (up to 100 %).

Fig. 5 Network topology for
WSN in COOJA simulation
environment

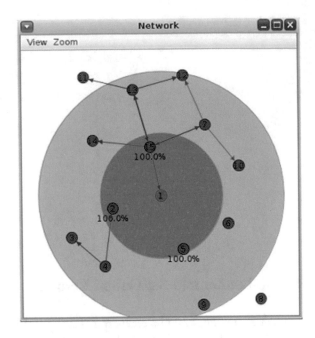

Table 2 General parameters for simulation

Parameter value	Value
Simulated mote platform	Tmote Sky
OS	Contiki
Chip	CC2420
Encryption and authentication	AES-CCM (counter with CBC-MAC)
Radio channel	UDGM
Bit rate	250 kbps
Node transmission range	50 m
Node carrier sensing range	100 m
Total frame size	127 B

As, in general, wireless sensor network protocols are highly data-dependent i.e. aware of the constraints on the transmitted data, this can be used to carefully configure the network in order to attain specified quantitative and qualitative performance indicators.

Security operations require limited flexibility in terms of RAM usage. There are also other factors that influence the resource consumption like the number of neighbors, transmission range or payload size. More processing more memory usage. In Fig. 6 the number of hops per node is illustrated. We need to keep these values in mind for the communications and security analysis at the individual node level.

Fig. 6 Network hops per node

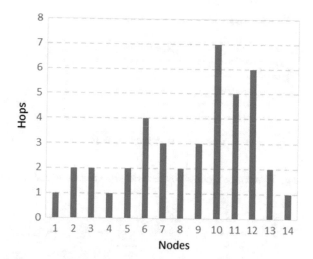

In an IEEE 802.15.4 system, all communication is based on packets. In order to evaluate the communication constraints of the proposed scheme in terms of total time consumed during the exchange of packets, we used the COOJA simulator. The simulation environment consists of one network as shown in Fig. 5. Each ContikiSec security mode has been evaluated and computational time and energy required for the encryption of IPv6 packets are offered in the subsequent figures.

We choose to run every experiment at 60 min simulation time. All data packets that are sent from the sender nodes throughout the 6LoWPAN network to the border router have a payload of 30 bytes. The radio is able to transmit a total of 48 bytes per data frame, in the ContikiSec-AE form. The payload is encrypted using the specific mode of operation presented in Table 1. We assume that all the nodes are provided in the initial setup with a single 128-bit key.

Figure 7 depicts the number of successfully transmitted/received packets of the three security modes examined. The results show that packet loss is higher for additional overhead like ContikiSec-AE has and also for the nodes that are multiple hops away from the sink node. These results show that ContikiSec-AE has an excellent throughput approximation.

Since energy is a limited resource in WSNs, we need to analyze the impact of the three security modes on the hardware performances. Contiki provides a software-based power profiling mechanism that keeps track of the energy expenditure of each sensor node. Being software-based, the mechanism allows power profiling at the network scale without any additional hardware.

In Fig. 8 is illustrated the average radio duty cycle per node, either in listen or transmit mode, computed as the time the processor is on. Due to number of hops, some nodes are acting as routers for their neighbors, therefore they have higher duty cycle and power draw.

As we can see in Fig. 8, the radio is on between 19.26 and 19.92 % of the time. The average activity time spent in radio transmitting and radio listening mode

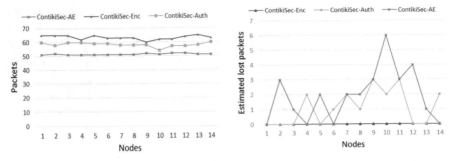

Fig. 7 Average successfully received (*left*) and estimated lost (*right*) packets per node

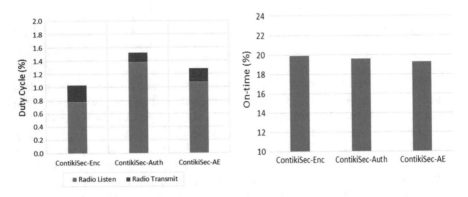

Fig. 8 Average radio duty cycle (*left*) and on-time activity (*right*)

ranges from 1 to 1.5 %. The total time is calculated as the sum of the time spent with the radio active and the time spent on listening mode. The numbers in Fig. 8 represent a metric for the energy-efficiency of different configurations, since they denote how much radio on-time has been spent per packet on average.

In low-power networks, the radio transceiver must be switched off as much as possible to save energy. This is strongly recommended for battery operated applications. In Contiki, low power consumption may be achieved by the Radio Duty Cycling (RDC) layer. The configured RDC and MAC protocols for our Contiki project are ContikiMAC and NullMAC. The 80 % radio sleep time is due to the ContikiMAC mechanism used in all simulations.

The power consumption is decomposed in categories, considering low-power memory, CPU and the radio listen and transmit modes. Power draw in listen mode is larger than in transmit mode as nodes spent most of their active time listening of packages from the sink, routing requests and route updates from neighbors than transmitting their own data (Fig. 9).

In the initial start-up phase, the client nodes have higher power draw, as the exchange of pairwise keys must be done, neighbor discovery, IPv6 addressing, etc.

Fig. 9 Simulated average power consumption

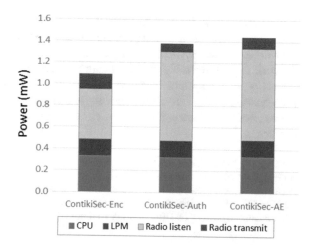

Energy consumption in WSNs may depend on the designed application and it is also hardware specific. Also, simulation results may vary from real hardware implementation. The ContikiSec header can represent a limiting factor on the maximum transmission rate of IPv6-enabled WSN applications. This factor, together with the impact of ContikiSec on the energy required from sensor nodes, allows us to conclude that this compressed security header should preferably be used with applications requiring high security.

6 Conclusion

In this paper, we have reviewed the security issues and two of most popular existing implementations/specifications for WSNs: TinySec and ContikiSec. Existing implementations of security primitives for WSNs have shown that designing security for WSNs is an interesting engineering problem, where specifics of WSNs should be considered. Our design was about performance analysis of security primitives using ContikiSec architecture, the evaluation was carried on the Tmote Sky platform. We provide simulation results and a perspective on security issues for this class of devices.

Even though receiving much attention recently, there are still many problems to be addressed in a WSN, such as discovery and revocation of compromised nodes or keys revocation from compromised nodes. Security mechanisms designed for low-power devices have the potential to adapt to the requisites of different WSN applications, while providing fundamental security independent of the application running on the sensor nodes. Our evaluation study demonstrated that this is possible with currently available limited sensor networks and contributes to definitions of new mechanisms that allow security integration on IPv6-enabled WSNs.

As future work, we plan to study the boundaries of using key management and also examine the effect of compromised nodes in secure network layer architectures for wireless sensor networks. Using a primitive key management scheme makes the network more vulnerable to various attacks. Introducing public-key cryptography protocols to the system (such as ECDH, ECDSA etc.) would make it fitter to IPv6-enabled WSNs. A clear necessity arises for experimental validation on real sensor nodes, following comparative experimental work on simulated mote platforms.

Acknowledgments The work has been partially funded by the Sectoral Operational Programme Human Resources Development 2007–2013 of the Ministry of European Funds through the Financial Agreement POSDRU/159/1.5/S/132397. This work was partially supported by a grant of the Romanian Space Agency, "Multisensory Robotic System for Aerial Monitoring of Critical Infrastructure Systems" (MUROS), project number 71/2013.

References

1. IEEE 802.15.4 Working Group: Part 15.4: low-rate wireless personal area networks (LR-WPANs), April 2012
2. Dunkels, A., Eriksson, N., Österlind, F., Tsiftes, N.: The Contiki OS—the operating system for the internet of things [Online]. Available http://www.contiki-os.org/
3. Halcu, I., Stamatescu, G., Stamatescu, I., Sgarciu, V.: An analysis of security and communication constraints of IPv6-based sensor networks. In: Proceedings of the 6th International Conference on Electronics, Computers and Artificial Intelligence (ECAI), 2014, pp. 55–60, 23–25 Oct 2014
4. Stamatescu, G., Sgarciu, V.: Evaluation of wireless sensor network monitoring for indoor spaces. Instrum. Meas. Sens. Netw. Autom. **1**, 107–111 (2012)
5. Shelby, Z., Bormann, C.: 6LoWPAN: The Wireless Embedded Internet. Wiley, London (2009)
6. Karlof, C., Sastry, N., Wagner, D.: TinySec: a link layer security architecture for wireless sensor networks. In: SenSys'04, 2004
7. Dargie, W., Poellabauer, C.: Fundamentals of Wireless Sensor Networks. Wiley, Chichester (2010)
8. Luk, M., Mezzour, G., Perrig, A., Gligor, V.: MiniSec: a secure sensor network communication architecture. In: Proceedings of the 6th International Symposium Information Processing Sensor Networks, pp. 479–488, April 2007
9. Jinwala, D., Patel, D., Dasgupta, K.: FlexiSec: A Configurable Link Layer Security Architecture for Wireless Sensor Networks, p. 22, March 2012
10. Casado, L., Tsigas, P.: Contikisec: a secure network layer for wireless sensor networks under the contiki operating system. In: Proceedings of the 14th Nordic Conference on Secure IT Systems: Identity and Privacy in the Internet Age, pp. 133–147 (2009)
11. Khan, S.U., Pastrone, C., Lavagno, L., Spirito, M.A.: An authentication and key establishment scheme for the IP-based wireless sensor networks. Procedia Comput. Sci. **10**, 1039–1045 (2012)
12. Xiao, Y., Rayi, V.K., Sun, B., Du, X., Hu, F., Galloway, M.: A survey of key management schemes in wireless sensor networks. Comput. Commun. **30**(11–12), 2314–2341 (2007)
13. Ferng, H.-W., Nurhakim, J., Horng, S.-J.: Key management protocol with end-to-end data security and key revocation for a multi-BS wireless sensor network. Wirel. Netw. **20**(4), 625–637 (2013)
14. Li, C.-T.: Security of wireless sensor networks: current status and key issues. In: Smart Wireless Sensor Networks, pp. 299–313 (2010)

15. Saad, L., Chauvenet, C., Tourancheau, B.: IPv6 routing protocol for low power and lossy sensor networks simulation studies. Sensors Transducers J. **14**, 79–92 (2012)
16. T.I. CC2420 Specification: 2.4 GHz IEEE802.15.4/ZigBee-Ready RF Transceiver. http://ti.com
17. Dunkels, A., Eriksson, J., Finne, N., Tsiftes, N.: Powertrace: network-level power profiling for low-power wireless networks low-power wireless. SICS Technical Report T2011:05. Swedish Institute of Computer Science (2011)
18. Eriksson, J., Österlind, F., Finne, N., Tsiftes, N., Dunkels, A., Voigt, T., Sauter, R., Marrón, P. J.: COOJA/MSPSim: interoperability testing for wireless sensor networks. In: Proceedings of the Second International ICST Conference on Simulation Tools and Techniques (Simutools 2009), pp. 27:1–27:7 (2009)

Index

© Springer International Publishing Switzerland 2016
E. Pricop and G. Stamatescu (eds.), *Recent Advances in Systems Safety and Security*,
Studies in Systems, Decision and Control 62, DOI 10.1007/978-3-319-32525-5

Printed in the United States
By Bookmasters